Charles Marion Tyler

The island world of the Pacific Ocean

Charles Marion Tyler

The island world of the Pacific Ocean

ISBN/EAN: 9783337101534

Printed in Europe, USA, Canada, Australia, Japan

Cover: Foto ©Andreas Hilbeck / pixelio.de

More available books at **www.hansebooks.com**

WAX-PALM OF BORNEO.

THE

ISLAND WORLD

OF THE

PACIFIC OCEAN

───────────

BY

CHARLES MARION TYLER

SAN FRANCISCO, CAL.:

SAMUEL CARSON & CO.

PUBLISHERS & BOOK-SELLERS,

3 SANSOME STREET.

1887

PREFACE.

Some years ago, it was the author's privilege to become interested, in a small way, in trade with the South Sea. Although not of great personal profit or benefit at the time, so many and varied were the commercial interests presented, that a journal was kept, of the ups and downs of trade and adventures among the Pacific Islands. These notes by the wayside, are merged in the following pages.

From the movements taking place among the great maritime powers of the world, England, Germany, France and America, in regard to the islands of the Pacific Ocean, with all of which the intelligent reader is familiar, the author is strengthened in the hope that a work relating to a region of such vast prospective benefits to the United States, and the world at large, may be read with some degree of interest as well as profit.

The immense field encompassed within the boundaries of Oceanica, together with many island groups lying beyond its limits, would make it almost impossible to visit and survey in the lifetime of any one person. From this fact, I am sure the writer will be dealt lightly with, for the necessary frequent reference made to the valuable writings, notes and personal experiences of others.

In the endeavor to brush away the cobwebs of time, or brighten up or throw light on the dark shadows cast over many portions of Oceanica by the veil of romance and tradition, I may perhaps be engaged in a herculean task.

The truth, however, I am sure will prove "stranger than fiction," and a good deal more profitable. In this respect some pains have been taken to make the following work in regard to history, discovery, ethnology, biography, chronology, geography, area, population and products, standard and reliable.

CONTENTS.

CHAPTER XI.

ISLANDS.

CHAPTER XII.

ISLAND PRODUCTS AND RESOURCES.

CHAPTER XIII.

ISLAND PRODUCTS AND RESOURCES.

CHAPTER XIV.

ISLAND PRODUCTS AND RESOURCES.

CHAPTER XV.

OCEANIC ETHNOGRAPHY.

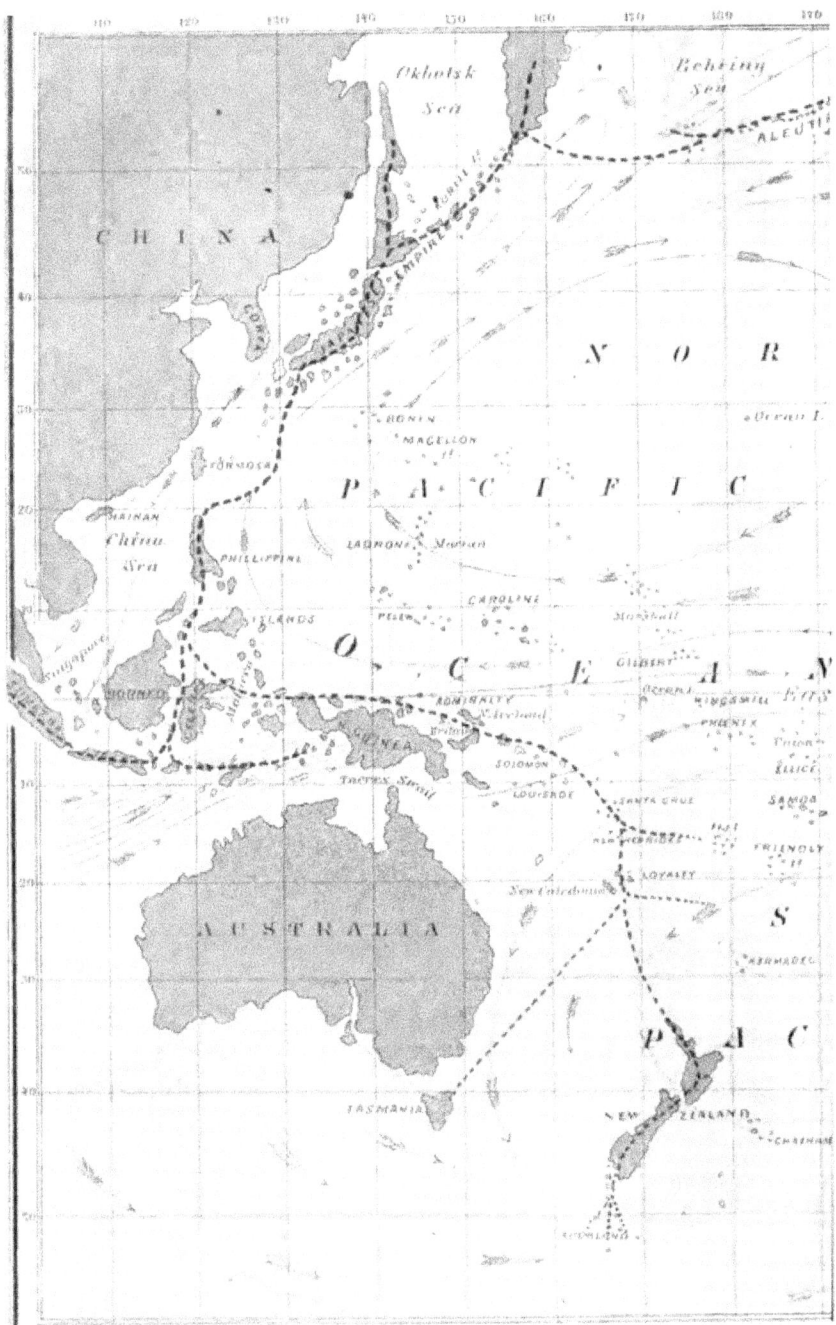

Okhotsk Sea

Behring Sea

ALEUTI

CHINA

EMPIRE

JAPAN

KURILE

NOR

Ocean I.

BONIN

MAGELLON

PACIFIC

FORMOSA

HAINAN

China Sea

LADRONE Mariana

CAROLINE

PHILLIPPINE

PELEW

Marshall

ISLANDS

Singapore

O

C

E

A

N

GILBERT

BORNEO

Molucca

ADMIRALTY

New Ireland

Britain

KINGSMILL

PHOENIX

Union

Ellice

GUINEA

SOLOMON

LOUISIADE

SANTA CRUZ

SAMOA

Torres Strait

NEW HEBRIDES

FRIENDLY

LOYALTY

AUSTRALIA

New Caledonia

S

KERMADEC

PAC

TASMANIA

NEW ZEALAND

CHATHAM

AUCKLAND

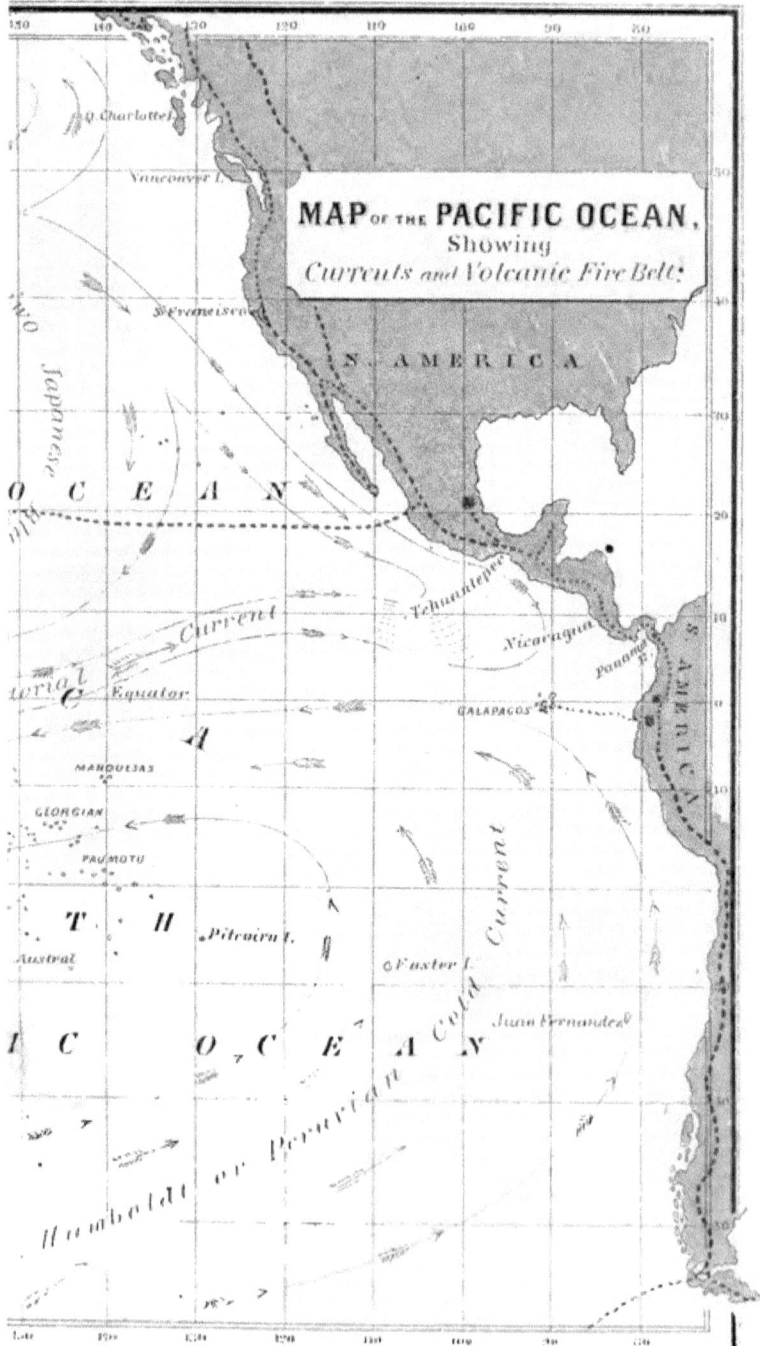

MAP OF THE PACIFIC OCEAN,
Showing
Currents and Volcanic Fire Belt.

N. AMERICA

S. AMERICA

Q. Charlotte I.

Vancouver I.

S. Francisco

Japanese

Current

O C E A N

Equatorial

Current

Equator

C

A

MARQUESAS

GEORGIAN

PAUMOTU

T H

Pitcairn I.

Austral

I C

O C E A N

Tehuantepec

Nicaragua

Panama B.

GALAPAGOS

E. Easter I.

Juan Fernandez

Humboldt or Peruvian Cold Current

CHAPTER I.

OCEAN LORE

Thou glorious mirror, where the Almighty's form
  Glasses itself in tempests; in all time,
Calm or convulsed—in breeze, or gale or storm,
  Icing the pole, or in the torrid clime
Dark heaving—boundless, endless and sublime.—
                    BYRON (*Childe Harold*).

PACIFIC OCEAN.

IN 1513, three hundred and seventy-one years ago,
Balboa was dragging the timbers of his ship across
the Isthmus of Darien, from the Atlantic to the
Pacific shore. Rebuilding his vessel there, he was the
first, in our modern day, to sail on the great ocean
waters. If gifted with supernatural vision, he would
have seen the Pacific Ocean, spread out over an area
of eighty million square miles, covering nearly all of
the western hemisphere. Its mighty waves, laving
the eastern shores of Asia on the one hand, and the
western coasts of the two great American continents
on the other. Reaching almost to the birth-places
of the ice-bergs of either pole, embracing the heat of
the Torrid Zones, it includes all the climates of the
world in its vast limits. He would have seen, north
of the equator, the Kuro Shiwo, the Japanese Black

9

Stream, sweeping in immense circles from left to right. South of the line, the Humboldt, or Peruvian Cold Current, circling from right to left. Both forming the highways over which it is thought the Asiatics voyaged to people our western world. In the depths of the great ocean—nearly three miles—almost beyond reach of the sounding-line, would be seen alike, the cradle and tomb of the island world of the Pacific.

Thousands on thousands of islands would come into view, like great emeralds dotting the mighty sea; with the tempest, typhoon and hurricane pursuing their furious course over the broad expanse of waters, subdued long before the transit of the great sea is performed—walled in and held back by the placid seas surrounding them.

So large, indeed, is the Pacific, greater in area than all other oceans combined, that the habitable portions of our globe, the land, would be lost in its limits, and yet a sea larger in extent than the Atlantic be left.

EARLY NAVIGATORS.

The discovery, location and conquest of many islands of the Pacific Ocean comes to us, out of the dim past, surrounded by a halo of romance. The names of famous navigators rise up in the mind, recalled by history, as pioneers in the mighty progress of the new world.

The quaint accounts of Captains Cook and Wallis were taken up and confirmed by the old salts of Nantucket and New Bedford. Their stories of the wealth, beauty and fertility of the myriads of islands met with in their whaling voyages, has long excited the curiosity of the world. The singular fatality, too, that seems to have followed nearly all the fathers of navigation in the Pacific has but added interest in their voyages and discoveries. Vasco Nunez de Balboa, who took possession of the entire South Sea in the name of the Pope, fell under the headsman's axe. Magellan, the first to reach the Indies, by a western route, through the Straits that bear his name, died by the sword, in a petty religious quarrel with some island king. Alonzo de Saavedra, he who attempted the passage of the North Pacific, from Manilla to Mexico, the reputed discoverer of New Guinea, which he named Tierra de Oro, met a lowly fate on the equator.

This same Saavedra, was probably the first to propose cutting a canal through the Isthmus of Darien, at Panama. In his proposition to the King of Spain on this subject, and in his memoirs, he goes into the enterprise in detail, and recommends the forcible employment of the inhabitants of that region, to accomplish his object. He states that "Providence had evidently placed them there, in order that they by their labors, might assist in the extension of the commerce of Christendom." Captain Cook fell among the savages of the Sandwich Islands. Sir Humphrey Gilbert was lost in a storm at sea. The chains and anchors of the vessel of M. de la Perouse were found, but his ultimate fate has never been ascertained. Others, like William Dampier, Roggewein and Fernando Quiros, were destined to what many would consider a more melancholy ending. Dying in poverty, forgotten, unhonored and unsung, in their native land. Of Quiros, Cardinal Valenza says: "I have seen in a wine-shop of Seville, one Fernando Quiros, who had been an adventurer in the Indies and beyond, and who told me he had seen there people who did eat their wives and other relatives, in place of consigning them to the tombs, which did not so much surprise me, seeing that the same thing has been related by the ancients."

It is only of very late years that we—I speak more particularly of the inhabitants of the Pacific Coast—have begun to reap commercial benefits, in a large way, from traffic with the islands of the South Sea. Growing rapidly as we are in wealth and population, the time has come when every effort should be made to encompass a large share of the trade. The wonderful impetus which is now being given to commercial enterprise on the western coast of North America, by the completion of three continental railways across our country, with still another under way, will go far to make San Francisco one of the greatest commercial cities on the globe.

Fifty years ago the multitudinous islands of the Pacific were but little known. Their vast number, nearly 8,000—their area fully 4,600,000 square miles —populated by over 77,000,000 inhabitants, are yet almost as an unknown land to our people.

If we compare the area, exports and imports of the Sandwich Islands (with the port of San Francisco alone) with the area of the Pacific Islands, whose exports and imports, are now about $700,000,000 per annum, the values would reach the vast sum of $7,790,-000,000. Something over five times the average annual exports and imports of the United States for the last four years.

Truly "there is a wonderful land; a land of fertility, of spices, of valuable fibres, of sago and cinnamon, of sandal wood and gold," and without a doubt the islands of the South Sea, are of this description. The climate is unsurpassed in any part of the world, and is both conducive to health and longevity. With the thermometer rarely below 65 degrees, and hardly ever reaching above 86 degrees, we have a perpetual summer of so delightful a temperature that working men of Europe or America may devote themselves to a life of pleasant and profitable labor all the year 'round. If we add to this, lands of inexhaustible fertility, we have within easy distance of our port millions on millions of acres of soil, far surpassing that of the famed West Indies.

When it is considered that the area of the latter islands is only about one-fiftieth of the islands of the South Sea, that the population exceeds that of the West Indies nearly eighteen times, that the exports and imports of the latter are over seventy-five millions per annum, some idea may be formed of the vast commercial interests that will arise from the occupation, development and trade with the islands of the Pacific.

With the exception of the more prominent islands put down on the list, already well known to the commercial world, the great majority remain as a sealed book, so far as their agricultural, mineral and other qualifications are concerned. In fact, if any trade has ever existed among them, it has been carried on by men without means, who have become tired of the sea, castaways, pirates and refugees. A class as much to be feared as the traditional man-eater. It is not strange, therefore, that the many valuable interests that could could be developed in these garden-spots of the world suffer and languish when under the control of such spirits.

THE JAPANESE BLACK STREAM.

San Francisco lies directly in the track of the great ocean current, that, like the Gulf Stream of the Atlantic, flows in the Pacific. This current is known as the Kuro Shiwo, or Japanese Black Stream. If we assume its point of commencement to be off the coast of Japan, it would trend northerly from that country, one portion flowing to and around the Aleutian Isles in Behring Sea, while the other, or main current, flows more to the east, towards our northern coast, which it reaches just south of Queen Charlotte Islands, off the coast of British Columbia. Running thence southerly along the shores of Washington Territory and Oregon, and along the coast of California, it turns south by west just off the harbor of San Francisco. Dividing again at this point, one stream flows by the Hawaiian Isles, and westerly among the islands of the North Pacific, and again northerly to the coast of Japan. The other division flows in a more southerly direction until the equator is reached, where it turns to the west, running among the myriads of islands in that region, turning to the north, and flowing by and to the east of the Philippines, reaching an assumed starting point off the coast of Japan. The Kuro Shiwo flows at the rate of ten to fifteen miles per day, and must in its great silent way render invaluable service in helping preserve the temperate climate along the coast as well as in the interior of the States bordering the Pacific. Its great value in favoring commerce to and from our port with the lands of the South Sea, can hardly be over estimated.

With the completion of the Panama and Nicaragua canals, great commercial gateways will be opened between the Pacific and Atlantic oceans, and a trade with the whole world offered to the islands of the Pacific that will in time become gigantic in its proportions. To encompass this, it is safe to predict that the great maratime powers will contend.

Commercial cities like New York, Boston, New Orleans and San Francisco will grasp at the facilities offered by the short routes created by the canals, and the sails of all nations will dot the southern seas.

A personal experience among the islands of the Pacific in commercial and other ventures, leads me to write on this subject, with a little knowledge and a great deal of interest. While it would take volumes to do justice to this subject, in a detailed account, it shall be my endeavor to present such features and facts in a condensed form as may interest and prove of value to many readers. Among the myriads of islands, which I have placed in groups, as will be seen in the tables, I have given but a short description of one or more in a group. It being understood that the description may be accepted as generally applicable to all the islands of a particular cluster, excepting perhaps, the number of inhabitants, size and locality.

SOUTH SEA BUBBLE.

One of the obstacles to be surmounted in favorably presenting the vast interests to be found in those garden-spots of the world, the Pacific Islands, is the ban put upon all concerted ventures that have been attempted in these regions by the great financial crash of the South Sea company in years long passed away. In fact, the term "South Sea Bubble" is generally used as a synonym for all enterprises not based upon solid foundations, the popular impression prevailing that the great failure of this company came from commercial and other ventures made in the South Sea.

The truth is, the company had no ventures or interests in that region resulting in failure, for if we except *one* vessel only, that made a trading voyage in 1717, and that, too, to Spanish South America, in the interests of the corporation, there are no accounts of practical commercial operations entered into in the Pacific by this company. True, they had some valuable privileges from the English Government, as well as from Spain, that theory and misrepresentation easily built into a supposed practical trading monopoly, although their operations were principally financial and stock jobbing, and confined altogether to London.

The fabulous stories and traditions of the Spanish South American countries, among which were Chile and Peru, the vast wealth in gold, silver and jewels, together with well concocted stories of the wonderful productions of the soil, and the supposed exclusive rights obtained from the king of Spain, formed the cornerstone of the South Sea Company. After the treaty of Utrecht, Spain withdrew all grants and privileges made to the corporation, yet the wealth and power of its directors, with the prestige of a long list of rich stockholders, enabled the company to retain a footing in great financial circles. As a valuable support to the schemes of the corporation, the wonderful products of the Pacific Islands, then making their way into all parts of Asia and Europe, were used as a lever in its advancement. The shells, pearls, fruits and spices, the whalebone and oil, the rich results of land and sea, were cunningly interwoven into a project that at one time set all Europe wild with greedy anticipation.

In 1711, the Earl of Oxford, who was Lord Treasurer of the Kingdom, finding the credit of the Government somewhat impaired, conceived the scheme of funding a portion of the national debt of Great Britain. then amounting, in round numbers, to $155,000,000. Of this sum, he proposed to fund $50,000,000 by issuing bonds of the Government, which were to be paid, interest and principal, by special regular duties upon silks, wines, tobacco, and some of the other most valuable importations. Purchasers of the bonds were to receive a certain amount of South Sea stock with each Government bond, that stock being then considered of sufficient value to offer a tempting bait in attempting to float the amount required. The credit of the Government, with the six per cent. interest, together with the shares of the South Sea Company, and certain trading privileges allowed to the corporation in trade with South America, made it an easy matter to fund the $50,000,000.

Meantime, the company was using every influence to establish and enlarge its credit, and though partially opposed in its schemes by many of the great statesmen and financiers of Europe, the Bank of England and the East India Company, the advancement of the "bubble" interests met with a curious success on every hand But it was not until 1720 that the company reached the zenith of its influence and power. which culminated in offering to take the whole national debt of Great Britain on its shoulders at a reduced interest, but otherwise on similar terms to the first loan. In 1719, so many and great had become the

schemes of the company, that it was found necessary
to increase its capital stock to nearly $60,000,000, with
shares set at a par value of $500.

The Bank of England, fearful of the rapidly-growing
power of the South Sea Company, made a similar
proposition to the Government, offering as a premium
$15,000,000. This offer was more than doubled
by the South Sea Company. Under the wing of even
royalty itself, and with emissaries and agents in every
quarter promulgatiug the most fabulous stories, backed
up by the free use of money and presents of stock, the
corporation had their offer accepted in both Houses of
Parliament, by a vote of 83 to 17 in the House of Lords
and 172 to 55 in the Commons. So well were the plans
laid, and so general was the desire for speculation, that
the shares of the company were eagerly sought after
at $1,500 per share. On the 14th day of April, 1720,
subscription books were opened to the public, of $10,-
000,000 of stock at $1,500 per share, and was almost
immediately taken, with $1,000,000 more before the
books were closed. On the 30th of April of the same
month and year, a further amount of $5,000,000 was
offered at $2,000 per share, and the amount taken in a
few days, and $2,500,000 in addition. As an illustra-
tion of greed and infatuation of a speculative people,
hoodwinked by stories only found in sober moments in
the "Arabian Nights" and tales of like ilk, history furnishes but few equals. Rich and poor alike parted
with the most substantial securities, many leaving them
in the hands of the company to secure a preference of
shares, without limit as to price. The stock rose rapidly to $2,500, $3,000, $3,500, with many fluctuations,
and reached the top figure of $5,000 per share, equal to
$300,000,000—when the bubble burst. It gradually

leaked out that the chairman of the company, Sir John Blunt, a man of low origin but extraordinary financial ability, and one of the chief projectors of the scheme, together with the favored few having the management of its affairs, were selling out. The ruin and desolation that followed—the disappointment, rage and desire for revenge of the deluded ones—turned all England into a chaos of financial distress.

Parliament was convened, and measures immediately taken for the punishment of the schemers, who, but a little while back, were lauded as the kings of finance. Many of the leaders were arrested and imprisoned, and a fine of $10,000,000 imposed and collected, to be distributed among the deluded stockholders. The Bank of England and the East India Company were induced to come to the rescue, they taking and sustaining millions, and easing down one of the greatest financial crashes in the history of any country. Enough of the stock and bonds of the company were secured, together with the fines imposed, to enable the Government to declare a dividend among the stockholders of nearly forty per cent., still leaving an immense sum to be carried and taken care of by the Government.

One hundred and twenty-five years after the incipiency of this scheme, I find the following in a financial statement of the funds of Great Britain :

South Sea Debt and Annuities.—This portion of the debt, amounting, on the 5th of January, 1836, to 10,144,584 pounds sterling, or $50,722,920 of our money, is all that now remains of the capital of the once famous, or rather infamous, South Sea Company. The company has, for a considerable time past, ceased to have anything to do with trade, so that the functions of the directors are wholly restricted to the transfer of the company's stock and the payment of the dividends on it, both of which operations are performed at the South Sea House, and not at the bank. The dividends of the old South Sea annuities are payable on the 5th of April and 10th of October; the dividends on the rest of the company's stock are payable on the 5th of January and 5th of July.

In 1727, three-fifths of the public debt of England was held by the South Sea Company—or about two hundred and seventeen millions, five hundred thousand dollars.

CHAPTER II.

ISLANDS

Call us not weeds, we are the flowers of the sea.
E. L. Aveline.

GALAPAGOS GROUP.

IN making a journey through these garden spots of
the Pacific, for geographical reasons, it is assumed
that our voyage commences at the Galapagos Islands; and that all longitudes are taken from Greenwich, east or west, as the case may be.

The Galapagos, some fifteen in number, lie on
both sides of the equator, being about 600 miles westerly from the coast of Ecuador, to which republic they
belong. Their area is 3,000 square miles, with a population of 4,000. The principal islands in the group are
Albemarle, James, Chatam, Indefatigable, Hood,
Charles and Narboro. Their curious geological formation, and evident volcanic origin, has given rise to
much speculation on the part of scientists. There are
to be seen in the group nearly 2,000 craters of extinct
volcanoes, leaving one with the impression, that a permanent residence here, with the fear of an eruption
continually before the mind, would not be pleasant.
There is probably no place in the world, where turtles

are so abundant, as at these islands. In their laying
season they literally swarm along the shores, and are
hunted and slaughtered by thousands. An establish-
ment or several of them, might be located here for
catching and canning turtle, that would no doubt prove
a great success, and is well worth the thought and enterprise of the commercial world.

THE MARQUESAS GROUP.

Leaving the Galapagos, we sail away west by
south for the Marquesas Archipelago, discovered by
Mendana in 1595. The islands in this group stand
high above the level of the sea, some of the mountain
peaks towering up in the clouds, while their steep and
rugged sides, sweep down in many places to the waters
edge.

They are thirty-five in number, situated between
latitudes 7 deg. 53 min. and 10 deg. 30 min. south, and
longitudes 138 deg. 43 min. and 140 deg. 44 min.
west. The area of the whole group is something like
1200 square miles, with a population of 20,000 people.

We found the landings here very difficult, and
were forced to lay off and on, quite a distance from
shore. Nuka-Hiva, the principal island, is about
eighteen miles long from east to west, and ten miles
wide. After several attempts we finally made a landing, and were very agreeably surprised at the great
beauty and fertility of the lands back from the coast.
Many of the valleys in the interior were one mass of
tropical foliage, with the huts of the natives peeping
here and there, from among the groves of cocoanut,
bread fruit and orange trees. The natives, although
kind and hospitable to our party to the last degree,

were in appearance anything but attractive. The men particularly, being tatooed in all the different fantastic styles of that art. At a short distance they had the appearance of being clad in chain armor, painted blue. The women are much fairer than the men, and only tatoo the face, with a few disfiguring spots on the lips. We saw several Polynesian Bibles in the huts of the natives, nearly all of whom claim to be Christians. Yet from all accounts we were among the decendants of veritable man eaters; people who practice all the heathenish and superstitious rites of their ancestors; and roast and eat their prisoners of war. Many of the islands of this group have well watered, beautiful valleys, well suited to the cultivation of coffee, sugar, cotton and other tropical products.

From the Marquesas we sail nearly due south, to that vast collection of coral islands known on maps and charts as the Low Archipelago or Paumotu Group. There are in all about seventy-eight islands and like the Marquesas and Society groups, are under a French protectorate. All except twenty of them are inhabited. The natives are a lawless and savage set, their greatest merit being the smallness of their numbers. However, some improvement has been noticeable among them lately, especially in their houses, clothing, and mode of living; the trade in pearls, pearl shell, and cocoanut oil, the principal products of this group, affording them the means for this desirable advancement.

ISLAND OF JUAN FERNANDEZ.

Still further south and to the east, in latitude 34 deg. about 400 miles west from Valparaiso, lies Juan Fernandez, in size some thirteen miles long by four miles wide, discovered in 1563 by the famous pilot and navigator, whose name it bears. It will always retain a marked prominence in island histories, being at one time the home of that celebrated, castaway Alexander Selkirk, whose life and adventures have been made so intensely interesting to youthful minds, and older ones too, for that matter, by Defoe in his wonderful book, "Robinson Crusoe." Selkirk was sailing master of the war galley *Cinque Porte*, and through a quarrel with Captain Straddling, asked to be put ashore on the island, which request was granted, and such supplies furnished him, as might be most needed in his lonely hermitage.

THE HOME OF CRUSOE.

In 1868 the officers of H. M. S. *Topaze* erected a tablet at the mouth of a small valley that traversed the land, and which gave the only clear outlook to the ocean from the island. At the northern end of this gap may be seen the tablet, with inscription reading: "In memory of Alexander Selkirk, mariner, a native of Laigo, in the county of Fife, Scotland, who was on this island in complete solitude for four years and four months. He was landed from the *Cinque Porte*, galley, 96 tons, 16 guns, in 1704, A. D., and was taken off by the *Duke*, privateer (Captain Wood Rogers), 12th of February, 1709. He died lieutenant of the *Weymouth*, in 1723, A. D., aged forty-seven years. This tablet was erected near 'Selkirk's Lookout,' by Commodore Powell and officers of H. M. S. *Topaze*, 1868, A. D."

In justice to the author of Crusoe, I quote still further, from the journal of the officers of H. M. S.

SCENE IN THE TROPICS.

Zealous: "We left Torne on December 21st, and arrived at the island of Juan Fernandez early in the morning of the 24th. It is difficult to imagine a more impressive bit of scenery than that which greets the eye on coming on deck, and seeing it for the first time after anchoring. We lay close to the shore, which went up almost perpendicular to a height, in some places, of 3,000 feet, towering above us like a huge giant. These heights faced us in the shape of a semi-circle, and to all appearances we lay in the middle of an extinct crater, of which the other half of the circle had been thrown into the sea, and now formed our anchorage. Every appearance justified this idea. No doubt a vast eruption took place many years ago, which produced this wonderful formation. At night particularly it looks very grand, and from its closeness and height, appears to be right over your head, standing out clear and distinct against the sky.

"The island belongs to Chili, and there are now resident on it five families, possessing nineteen children, three cows, four sheep, several horses, and goats innumerable, which latter abound on the other side of the island. The principal personage in this little community spoke English remarkably well. He told us they were perfectly happy, never were ill, and had no desire to leave the island. A state of bliss comprised in these three statements difficult to be understood; but though only attributable to the lowered state of the intellectual faculties, a state which it would be good to meet with more frequently amongst cultivated nations. Juan Fernandez was discovered in 1567, but from that time, I should imagine, no advantage was taken of its discovery—except occasional visits of buccaneers—till the year 1705, when

Alexander Selkirk was placed on shore for mutiny towards his captain. For more than four years he lived alone on this island, when at last he was discovered and taken off by Captain Rodgers, amongst whose crew was a man who had been on board Selkirk's ship when he was put ashore. From Selkirk's narrative Defoe is said to have derived and written his wonderful book, 'Robinson Crusoe.' Whether he did so or not, has been the subject of much controversy. I will not attempt to lay a dictum, for I do not think it matters now in the slightest either way. But in the memory of Defoe, who, as a writer, has had few equals before or since, and for the benefit of any one interested in the question, I must say that, having been led in the imagination to picture this island somewhat according to the book, there is nothing in Juan Fernandez to give rise to the belief that Defoe could have received from Selkirk anything but the idea from which he constructed his famous romance. Moreover, it was not published till the year 1719, ten years after the return of Selkirk.

"That Defoe took the greater part—as he has been accused—of his story from Selkirk's journal, it is impossible for anyone who has seen the Island of Juan Fernandez to believe. His cave can be seen now, cut in a sand-cliff, with the shelves in it used for cooking utensils, etc.; so that, unless we concede the almost impossible theory that when it was visited by a fearful earthquake, in 1760, the whole island changed its nature and appearance, we must acquit Defoe of plagiarism. If he did read Selkirk's journal, it had the effect simply of making him strive in every way to show there was no connection or similitude, the one with the other."

PITCAIRN ISLAND.

This little dot on the great ocean's surface, lying in latitude 25 deg. 3 min. south, and longitude 130 deg. 6 min. west, is about 2½ miles long by 1¼ wide, made famous as the home of the mutineers of the ship *Bounty*. It has, in addition, been of great service to the maritime world, being one of the fresh-water stations resorted to by whalers and others sailing in the Pacific.

Pitcairn Island was discovered in 1767 by Philip Carteret, navigator, who first sailed under Captain Wallis in 1766.

Although the history of the *Bounty* mutineers has already formed the theme of numerous writers, a very brief statement of the facts may not be out of place here, and might prove interesting to the general reader. Captain Cook, in his first voyage to Tahiti, one of the Society group, was much pleased with the bread-fruit tree, found in great abundance there, and on his return suggested to the British Government its many valuable qualities, not only for the nutritive uses, as food, of the fruit, but for the value of its timber and bark in a commercial way. He suggested transplanting the young shoots of the tree to the West India Islands, and the vessel *Bounty* was dispatched to Tahiti for this purpose, under command of Lieut. Bligh. It was during the voyage from Tahiti, loaded with the plants, that the mutiny occurred, Bligh being set adrift in an open boat. The mutineers returned to Tahiti, where they remained some time, recruiting their forces with natives—also persuading some of the gentler sex to accompany them —when they sailed away, reaching Pitcairn Island in 1789. There they established a colony, and after

using everything of value belonging to the ship for
building and other purposes, the *Bounty* was burned.
Many years elapsed before they were discovered, and
then only by accident, through an American ship captain who landed there for water. This being communicated to the British Government, a vessel was sent,
not only for their relief, but to punish the ringleaders
of the mutiny.

Lieut. Bligh, after many adventures and hair-
breadth escapes as a castaway, finally succeeded in
getting back to England.· He was placed in command
of another vessel, and successfully accomplished the
object of his first voyage, transplanting the bread-fruit
tree of the South Seas to the West India Islands in
1792–3.

BREAD-FRUIT TREE.

The bread-fruit tree (*Artocarpus incisa*) alluded to
above, is indigenous to nearly all the islands of the
South Sea, forming, with the cocoanut and banana,
the principal sources of food for the indolent natives.
The tree grows from twenty to forty feet high, with a
diameter of one to two feet. The bark and inner portions furnish a valuable fibre, while the pith supplies
the material for much of the paper cloth worn by the
natives.

The fruit ripens at different periods of the year.
It is about the size of a melon, and is found singly and
in clusters attached to the branches of the tree. There
are two or three periods in its growth when it can be
used ; at one time supplying a milky nutritious fluid as
a drink, and at another a delicious custard, but the
period when it is most used is just before ripening, at
which time the fruit is picked and baked in rude ovens,

the whole interior, assuming the spongy ·form of freshly-baked bread, with a pleasant taste—much superior to the doughy preparations, called bread, so common in Europe and America. When baked in this way. the bread-fruit can be kept for several months.

The timber of the tree is used to make many articles of furniture, and the trunk often formed into canoes, etc.

ISLAND GROUPS OF THE PACIFIC OCEAN.

NAMES OF GROUPS.	NO. OF ISL'S	LOCATED BETWEEN LATITUDES. Deg. Min.	Deg. Min.	LOCATED BETWEEN LONGITUDES. Deg. Min.	Deg. Min.	AREA IN SQUARE MILES.	POPULATION.	GOVERNMENT.
NORTH PACIFIC.								
Sandwich Islands	10	18 54 N	23 34 N	154 50 W	161 55 W	6,000	65,000	Native Kingdom.
Ladrone	20	13 10 N	18 50 N	144 40 E	146 3 E	1,254	6,000	Spain.
Pelew	20	6 46 N	8 11 N	134 10 E	134 45 E	905	10,000	Native, United States, etc
Marshall	32	4 10 N	13 50 N	165 15 E	172 10 E	1,000	8,000	U. S. Protectorate.
Caroline	500	4 30 N	19 50 N	137 10 E	163 15 E	15,000	38,000	U. S. Protectorate.
Phillippine	1200	5 15 N	19 44 N	119 45 E	126 31 E	120,000	6,500,000	Spain.
Celebes	1	2 0 N	5 45 N	118 45 E	125 10 E	72,000	2,000,000	Dutch, etc.
Galapagos	15	1 35 S	0 45 N	89 30 W	91 45 W	3,000	4,000	Ecuador.
Molluccas	100	3 0 N	5 45 S	124 0 E	130 0 E	10,000	150,000	Dutch, etc.
Java	30	6 0 S	8 45 S	105 5 E	114 31 E	52,000	18,000,000	Dutch, etc.
Little Java	1	1 0 S	6 55 S	105 5 E	114 31 E	4,000	800,000	Dutch, etc.
Borneo	15	4 10 S	6 10 N	108 45 E	118 33 E	286,000	2,184,000	Dutch, etc.
Sumatra	20	5 30 N	6 10 S	95 15 E	106 0 E	160,000	4,500,000	Dutch, etc.
New Guinea	25	0 45 S	10 45 S	130 30 E	151 10 E	300,000	650,000	Dutch, English, German.
Aleutian Isles	60	51 20 N	55 40 N	160 0 W	172 25 E	8,000	5,000	United States.
Bonin	70	26 30 N	27 45 N	140 50 E	142 30 E	650	2,000	England.
Japanese Groups	4000	26 10 N	54 40 N	126 40 E	151 30 E	250,000	34,000,000	Japanese Empire.
Sangir	50	2 0 N	4 N	123 0 E	125 0 E	2,000	30,000	Dutch, etc.
Chinese Groups	40	Lying near the Coast.				35,000	4,500,000	Chinese Empire.
British Am. Groups	25	Lying near the Coast.				21,600	30,000	England.
California Groups	15	Lying near the Coast.				600	500	Private Property.
Mexico (West Coast)	30	Lying near the Coast.				1,500	3,000	Mexico.

SOUTH PACIFIC.																Area	Pop.	
Admiralty	30	1	50	S	2	40	S	146	20	E	148	40	E	1,000	25,000	Germany.
Phœnix	8	2	40	S	4	45	S	170	30	W	174	40	W	300	1,500	U. S. Protectorate.
New Ireland	6	2	20	S	4	40	S	149	55	E	153	10	E	4,300	16,000	Germany.
New Britain	8	3	55	S	6	24	S	147	40	E	152	15	E	10,500	20,000	Germany.
Solomon	30	5	0	N	11	0	S	154	30	E	162	40	E	7,600	30,000	Independent, etc.
Ellice, Union, etc	100	5	29	S	8	50	S	176	10	W	179	10	W	1,160	8,000	U. S. Protectorate.
Marquesas	35	7	45	S	10	20	S	138	55	W	140	30	W	1,200	20,000	France.
Santa Cruz	20	9	40	S	11	50	S	165	50	E	167	10	E	400	7,000	France.
Louisade	40	10	30	S	11	40	S	151	30	E	154	10	E	1,200	5,000	Native, etc.
Samoa (Navigator)	10	13	30	S	14	20	S	169	22	W	172	50	W	1,650	40,000	Native, U. S., Germany.
New Hebrides	12	13	30	S	20	05	S	166	35	E	169	5	E	5,700	60,000	French, etc.
Paumotu (Low Arch.)	78	14	05	S	19	45	S	136	20	W	148	50	W	3,300	10,000	Independent.
Viti (Fiji)	250	16	0	S	20	0	S	176	42	E	178	30	E	7,400	120,000	Indep'dent, English, etc.
New Caledonia	1	20	0	S	22	50	S	164	0	E	167	45	E	6,000	60,000	France.
Tonga (Friendly)	100	18	22	S	21	50	S	173	40	W	175	24	W	1,000	25,060	Native.
Cook's	10	18	50	S	22	05	S	154	40	W	160	40	W	300	8,000	England, etc.
Loyalty	8	20	10	S	21	40	S	166	20	E	168	5	E	2,500	18,000	Independent.
Society	10	16	10	S	17	50	S	149	15	W	152	0	W	600	20,000	French Protectorate.
Banks	8	13	10	S	14	0	S	166	30	E	167	10	E	300	6,000	Independent.
Australia	82	10	50	S	39	10	S	113	0	E	154	0	E	3,000,000	2,000,000	England.
Tasmania	17	40	40	S	43	38	S	143	50	E	148	20	E	22,629	110,000	England.
New Zealand	16	34	12	S	47	18	S	166	25	E	178	30	E	122,582	476,000	England.
W. Coast S. America	300	Lying near the Coast.												35,000	30,000	Chile and Peru.
Austral	20	22	0	S	28	0	S	143	0	W	153	0	W	1,500	3,000	Independent.
Kermadec	6	30	0	S	33	0	S	178	50	W	180	0	W	500	1,000	Independent.
Gilbert	16	3	20	N	2	40	S	172	30	E	177	15	E	800	25,000	U. S. Protectorate.
Miscellaneous	500	Islands and atolls. (Estimated.)												42,000	100,000	Various Nations.
Total	8000													4,631,330	76,730,060	

CHAPTER III.

Behold the threaden sails,
Borne with the invisible and creeping wind,
Draw the huge bottoms through the furrow'd sea,
Breasting the lofty surge.

(*Henry V. Act III.*)

AUSTRAL ISLES.

WEST by north from Pitcairn, and almost due south from the Paumotus, lie the Austral Isles. The group, fifteen or twenty in number, are between latitudes 22 deg. and 28 deg. south, and 143 deg. to 153 deg. west longitude. The islands are small, and of but little commercial value at present. Rumbia, Tubuaia, Vantaia, Rumbaia, Bapai, Nelson and Oparo are the largest and best known of the group.

GAMBIER GROUP.

Another island cluster, the Gambier, due south from the Paumotus, are rapidly growing in commercial importance. The products, similar to those of the Austral Isles, are altogether of the tropical kind ; the soil rich and productive, well suited for the cultivation of coffee, cotton, sugar and spices. It is not my purpose to describe island groups located like the Austral and Gambier, in more than a general way. Lying, as they do, on the outside of the present valuable portion of the island world, their value is in the future.

SOCIETY ISLANDS.

Prof. Dana in speaking of this group says "that they consist of ten islands, ranging in a line 250 miles long trending N. 62 deg. W. Commencing from the north west they are as follows: Tubuai, Maurua, Borabora, Tahaa, Raitea, Hauhine, Tapuaemanu, Eimeo, Tetuaroa, Tahiti. To this number Osnaburgh or Metia, may properly be added, as it lies in the same range, about one hundred miles to the westward of Tahiti. With the exception of Tubuai and Tetuaroa, they are all basaltic or high islands. The area of the whole does not exceed twenty-five miles square, or 600 miles, and of these about one-half, or three hundred square miles, belong to the single island of Tahiti.

"These basaltic islands are characterized by high mountains, deep precipitous gorges, and that rich livery of green with which the mild airs of a perpetual summer clothe the tropical islands of the Pacific. Coral reefs in some instances border their shores, forming a circle around, dotted with verdant islets.

"The broken character of the surface is most striking on Eimeo, yet all the islands afford scenes of grandeur unsurpassed in the Pacific. In the distant view, Eimeo seems to be a mass of mountain towers, crags and peaks, rising abruptly to great elevations,

and in one lofty summit, resembling a rudely shaped
cone, there is a hole opening through, a few hundred
feet from the top. On Tahiti, still loftier summits,
with crowns and crests and jagged ridges constitute
the surface. The eye follows up one precipitious slope
to plunge at once one or two thousand feet to the
bottom of another.

"The islands to the north-westward are described
as exceeding Tahiti in their bold features, and in the
indentations of their shores, which form deep bays,
penetrating far among the mountains; they are for
their size, the most remarkable in the Pacific. There
is great luxuriance of verdure over the Society Islands,
and good soil. But owing to the mountainous character of the lands, and especially the remarkably steep
declivities, but little of the surface, comparatively, can
be brought under cultivation. Yet there are many
fine valleys, besides the level areas along the shores
which might be tilled to great advantage. The sugar
cane and many tropical fruits are already grown in
abundance, and to these the coffee plant and other
productions of the East Indies might be added."

TAHITI.

Having cargo for Tahiti, it was our good fortune
to remain several days, and of course time for a partial inspection of what has been so much written about.
The entrance to the main harbor of Tahiti is rather
difficult to navigate, and requires the assistance of
some ancient weather-beaten mariner who knows every
foot of the channel from boyhood. They are to be
found among the natives, who, for a proper consideration, will place your vessel at safe anchorage in the

inner harbor. The trade of these islands with the outside world is considerable, the exports reaching a value
of nearly one million of dollars annually, with imports
of as much more. Coffee, cotton and sugar-cane, as
well as all other tropical plants, do well in the group,
giving not only employment to the natives, but many
who are brought from other islands and China. The
people are intelligent and kindly disposed, and the
stranger may revel in all the delights of a tropical climate without let or hindrance. Missionary schools are
to be met with on nearly all of the isles, and the strict
observance of laws, as customary in our own country,
is enforced by the Government. Tahiti, although of
wonderful fertility, and better known to the world, has
many rivals in extent and rich soil; notably the islands
of Raitea and Huahine—both of the Society group—
where can be found beautiful valleys, with an abundance of water and a luxuriant vegetation of nearly all
the tropical fruits, which clothe the valleys, hills and
mountain sides to their very tops. Much could be
written of Tahiti that would prove interesting to the
lovers of curious traditions, and a great deal might be
said of Captain Cook and his voyage to this island—
sent by the English Government to take observations
of the transit of Venus. The shade of the tamarind
tree planted by Cook may be enjoyed, and relics from
the observatory built by himself and companions can
be carried away in quantities to suit. But space will
not permit many details in a subject so vast as the
Islands of the Pacific.

TONGA, OR FRIENDLY ISLANDS.

To the south and west of the Society group lie
the Tonga or Friendly islands, nearly one hundred in

number, and, like nearly all isles in this region, are formed on the coral reefs. The archipelago is divided into several groups—Tongatabu, Namuka, Hapai and Katoo being the largest and best known. The islands are very low, the highest ground seldom rising above an altitude of 100 feet. The products are similar to those already described; the natives are peaceable and friendly, nearly all of them professing Christianity.

The number in the group I have placed at one hundred; some authorities state as high as one hundred and fifty; with a total landed area of but 1000 square miles. They were discovered by Tasman in 1643, and visited by Captain Cook many years afterward, who gave them the name they bear to-day.

Of the inhabitants, it is said that they "are intellectually, perhaps, the most advanced of the Polynesian race, and exercise an influence over distant neighbors, especially in Fiji, quite out of proportion to their numbers, which do not exceed twenty or twenty-five thousand. Their conquests have extended as far as Niue, or Savage Island, 200 miles to the east, and to various other islands to the north. In Cook's time, Ponlaho, the principal chief, considered Samoa to be within his dominions. This pre-eminence may, perhaps, be due to an early infusion of Fijian blood. Pritchard (*Polynesian Reminiscences*) observed such crosses to be always more vigorous than the pure races in these islands, and this influence seems also traceable in the Tongan dialect, and appears to have been partially transmitted thence to the Samoan. Various customs, traditions and names of places point to a former relation with Fiji, but Fijian influence in Tonga is insignificant, compared with that of Tonga in Fiji. Their prior conversion to Christianity gave the people material as well as moral advantages over their neighbors, and King George, a very remarkable man, and far in advance of his people, has, during a long reign, made the most of these.

"Agriculture, which is well understood, is the chief industry. They are bold and skillful sailors and fishermen ; other trades, as boat and house building, carving, cooking, net and mat making, are usually hereditary. Their houses are slightly built, but the surrounding ground and roads are laid out with great care and taste.

"There are some ancient stone remains here, as in the Caroline Islands, burial places (*feitoka*) built with great blocks, and a remarkable monument consisting of two large blocks with a transverse one, containing a circular basin in the centre.

"The chief articles of export are cocoanut-oil and copra, a little sugar, cotton and coffee, the cultivation of which is encouraged by the king, and fresh provisions for ships, as yams, pigs and poultry. The chief imports are cloth, cotton prints, hardware, mirrors, etc."

HERVEY OR COOK'S ISLANDS.

A little to the north and east of the Tongas are the Hervey or Cook's Islands ; Mangaia, Raratonga, Autaluke and Hervey being the largest. They are all of considerable commercial value, not only on account of their agricultural products, but for the great number of turtles and quantity of beche de mer taken in this group. Their products are coffee, cotton, sugar, tobacco, cocoanuts, oil, fungus, tomano wood and bananas. Nearly all the natives of this group can read and write, and profess the Protestant religion.

A great deal of time and money has been spent in this region, educating and reclaiming the heathen. It is lamentable though that in adopting our more civilized manners and habits, that the good and bad of our civilization could not have been separated. Many of the natives here, as well as among other groups of the Pacific, seem to take to the bad naturally, and in this particular locality it resulted in almost decimating the population.

Raratonga stands high above the sea level, nearly 3,000 feet, and the rich tropical vegetation covers the mountain sides clear to their summits. Streams of pure water flow through its valleys of rich alluvial soil, and highly cultivated plantations are to be met with on every hand. The inhabitants offer a pleasing contrast to some already cited, being a happy, peaceful and industrious race, in a comparatively advanced state of civilization.

FIJI ISLANDS.

Nearly due west from Cook's Islands we come to the great group of Viti, popularly known as the Fijis. They are 250 in number, with an area of some 7,400 square miles, and population of about 120,000. It is said that "a few islands in the northeast of the group were first seen by Tasman in 1643. The southernmost of the group, Turtle Island, was discovered by Cook in 1773. Bligh visited them in 1789, and Captain Wilson, of the *Duff*, in 1797. In 1827 D'Urville, in the *Astrolabe*, surveyed them much more accurately, but the first thorough survey was that of the United States Exploring Expedition in 1840." The group was annexed by Great Britain in 1874, and if not justly territory of that country, is practically under

the protectorate of England to-day. Situated in both
longitudes, that is lying either side of the meridian of
Greenwich, and between latitudes 15 deg. 42 min. and
19 deg. 48 min. south, in the track of much of our commercial trade with Australia and islands further west, the
Fijis are rapidly growing in commercial importance.
They offer a curious study of the past and present.
At one time, and that, too, within the memory of the
living, the Fijis were inhabited by a race of fierce and
warlike man-eaters, whose victims were roasted and
eaten, after undergoing all the hideous rites and
tortures that their savage natures could suggest.
Now the abode of peace and plenty, with churches,
schools and manufactures throughout the land. If I
mistake not, there are at present 1,400 schools and
200 churches among these islands.

The rapid advance made by the natives in civilization, in the arts and agriculture has made of these
once inhospitable shores a pleasant home and resort
for people of all nations.

The main islands are known as Viti Lavu, Van-
nua Lavu, Moala, Kiro, Lotia, Vunie, Kandavau,
Vatata, Valava Ovalau, Lakeruba, Vanua and Yasawa.
Mr. Consul March, in his report speaking of the
capabilities of Fiji, says: "The productions and resources of Fiji have been described in previous reports; it is sufficient, therefore, to state that these
islands, rich and fertile, yield an almost endless variety
of vegetable treasures. They abound in edible roots,
medical plants, scents and perfumes, and timber of
various descriptions; whilst sugar, coffee and tobacco grow most luxuriantly, and if cultivated, would,
I think, prove as remunerative as cotton."

The group, generally speaking, may be of vol-

TAHITI—SOCIETY GROUP.

canic origin, many evidences of igneous creation prevailing through most of the islands, with traces of extinct craters, whose ancient fires were probably quenched by the waters of the surrounding seas. On some, traces of the sedimentary formations are met with, while on others coral is found, a thousand feet above the ocean level, forced up from the depths of the sea. Taken in all, the physical configuration is hilly and mountainous, some of the crests rising to a height of four or five thousand feet. Blessed with an even temperature and an abundant rainfall, the valleys and slopes covered with verdure and forests in all stages of bloom and growth, a view of the group from the sea is extremely pleasing to the eye. Small streams flow through the valleys, some of them reaching the dignity of navigable rivers, with valuable agricultural lands to be found on the lowlands along their banks, that a little skill and energy, surely arriving with the strangers making their homes in the group, will develop into agricultural wealth. Then rice, sugar, coffee and cotton will vie with the natural products, the cocoanut, bread-fruit, banana, lemon and orange.

NEW HEBRIDES GROUP.

Lying farther west and a little to the south of the Fijis, are the New Hebrides Islands, twelve in number, the largest and best known being named Aneteum, Tana, Vate, Api, Aurora, Whitsun and Espiritu Santo. The last named, the largest of the group, is about 65 miles long by 35 wide. Inhabiting most of the islands may be found a people the most treacherous and quarrelsome in the whole Pacific. Lieut. Meade, R. N.,

who visited there in 1865, in describing Tana, and
which may be accepted as about their present condition, says : "Tana is about 25 miles long by 12 broad,
the population being between fifteen and twenty thousand. But since the introduction of European diseases
and weapons, there has been a steady decrease. In
1861 a third of the people died of the measles. The
state of morals is extremely low; the natives assert
that the present excessive licentiousness was introduced
by the whites, who formerly resided on the island.
The chiefs endeavor to get drunk every night on *kava*.
The women do all the work ; the men all the fighting,
which is their constant employment. Cannibalism is
the custom all over the island." In 1842 the bark
Rose, from Nantucket, engaged in whaling in these
latitudes, took as passengers twelve native missionaries, who had been educated and raised in Christianity
on some of the more civilized groups. These missionaries were sent to Tana as an experiment, and in the
hope of retrieving a fallen race. Arriving off the island,
a whale-boat was lowered and manned with a well-
armed crew, in addition to the twelve Christian workers. The crew were cautioned as to the treachery and
brutality of the natives, and on no account to make a
landing longer than just necessary to place the missionaries ashore. On arriving at the beach, the natives
swarmed to the boat and assisted the landing of the
religious workers with every show of kindness and
affection. Acting under strict orders, the crew of the
whale-boat immediately put back for the ship, and were
not three hundred yards from the beach when the natives fell upon the missionaries, killing them all in the
most barbarous manner, and in full view of the occu-
pants of the boat.

LOYALTY ISLANDS.

South and westerly from the New Hebrides we come to the Loyalty Islands, said to have been discovered by Captain Cook in 1774. The group is "about 60 miles east of New Caledonia, consisting of Uvea or Uea (the northernmost), Lifu, Toka, and several small islands, and Mare or Neugone. They are coral islands, of comparatively recent elevation, and in no place rise more than 250 feet above the sea. Lifu, the largest, is about 50 miles in length by 25 in breadth. Enough of its rocky surface is covered with a thin coating of soil to enable the natives to grow yams, taro, bananas, etc., for their support; cotton thrives well, and has even been exported in small quantities, but there is no space available for its cultivation on any considerable scale. Fresh water, rising and falling with the tide, is found in certain large caverns, and, in fact, by sinking to the sea-level, a supply may be obtained in any part of the island. The population, about seven thousand, is on the decrease. The island called Neugone by the natives, and Mare by the inhabitants of the Isle of Pines, is about eighty miles in circumference, and contains about six thousand souls. Uvea, the most recent part of the group, consists of a circle of about twenty islets, inclosing a lagoon twenty miles in width; the largest is about thirty miles in length, and in some places three miles wide, and the next largest is about twelve miles in length. The inhabitants, numbering about twenty-five hundred, export considerable quantities of cocoanut-oil. The Loyalty Islanders are classed as Melanesian; the several islands have each its separate language, and in Uvea the one tribe uses a Samoan, and the other a New Hebridean form

44   *THE ISLAND WORLD*

of speech. Captain Cook passed to the east of New
Caledonia without observing the Loyalty group, but it
was discovered soon afterwards, and Dumont D'Urville laid down the several islands in his chart.   For
many years after their discovery the natives had a bad
repute as dangerous cannibals.   Christianity was introduced into Mare by native teachers from Rarotonga
and Samoa ; missionaries were settled by the London
Missionary Society at Mare in 1854, at Lifu in 1859,
and at Uvea in 1865.   Roman Catholic missionaries
also arrived from New Caledonia, and in 1864 the
French, considering the islands a dependency of that
colony, formally instituted a commandant."

(Encyclo. Brit., vol. 15; Gill: Gems from the Coral Islands, 1871 ;
Macfarlane : Story of the Lifu Mission, 1873.)

NEW CALEDONIA.

New Caledonia with an area of 6,000 square miles
and a population of nearly 60,000, was discovered in
1774 by Captain Cook.   One of the most beautiful
and valuable islands in the South Pacific, has been
rendered almost valueless, by its appropriation in 1853
by the French, and since used by that government as
a convict settlement.   It differs materially from the
coral formations underlying many of the Pacific isles,
springing evidently from the older geological periods.
It is one confused mass of rocks, hills and mountains,
corrugated with beautiful valleys and running streams.
The hills and mountains are covered with forests of
fine timber, while an abundant natural growth of
nearly all of the tropical fruits, afford easy sustenance
to the not over industrious natives.   Noumea the

capital is in the southern portion, and has a fine harbor, that should be used for anything, but the wants of the scapegraces of France. Copper, nickle and cobalt are found in paying quantities, and very lately some important discoveries of gold have been made.

CHAPTER IV.

ISLANDS

The turf looks green where the breakers rolled;
O'er the whirlpool ripens the rind of gold;
The sea-snatched isle is the home of men,
And mountains exult where the wave hath been.

LYDIA H. SIGOURNEY.

MARSHALL ISLANDS.

THE Marshall Archipelago, consists of two nearly parallel chains of Atolls, from 100 to 300 miles apart, the west known as Ralik, the east as Radek chains. They are between 4 deg. 30 min. and 12 deg. N., and between 165 deg. 15 min. and 172 deg. 15 min. E., and run N. N. W. and S. S. E. They were discovered by Alonzo de Saavedra, in 1529, who observing the fine tatooing of the natives, (the first allusion to that practice in the Pacific,) called them Los Pintados.

Among modern voyagers, Wallis first visited them in 1767. Captains Marshall and Gilbert reached them in 1788, and Kotzebue in 1816, explored them more thoroughly. The east group contains fifteen or sixteen atolls, which range from two to fifty miles in circumference.

There is a curious tradition on the Liban island, of

the Darwinian fact, that the atoll, once formed the barrier reef of an island now sunk beneath the lagoon.

GILBERT ISLANDS.

The Gilbert Archipelago, discovered by Com. Byron in 1765, is geographically, a south continuation of the Marshalls, the channel separating them being about 150 miles wide.

Several of the islands have good anchorages inside of the lagoons, with entrances on the lee side. On some the lee or west reef is wanting, owing to the abrading force of the west storms. During these, large trees, are washed ashore, their roots containing pieces of fine basalt, of which implements are made. There is a larger proportion of land to submerged reef and lagoon than in the Marshalls; the land sometimes rising twenty feet above the sea, whereas in the Marshalls the average level of the reef rocks above the surface is less than one foot; but, though the supply of fresh water is great, in fact enough for the luxury of a bath, the soil, especially in the south, is much less productive; yet the population is very dense. The usually scattered houses are replaced by compact rows of roofs, shaded by the cocoa palm, and, each with its boat shed below, line the shore.

Their number may be set down at sixteen, lying on both sides of the equator between 3 deg. 20 min. N., and 2 deg. 40 min. S. latitude, and 172 deg. 30 min., and 177 deg. 15 min. E. longitudes, with a landed area of 800 square miles and a population of 25,000.

These atolls may contain a greater number of people than mentioned, as the population seems very dense. This is accounted for by the small width of

the atolls, ranging from a few hundred yards wide
only, in some places, to several miles in others, and
the habit of the natives of flocking or swarming from
one island to another, or to particular localities on one
island. This occurs sometimes twice in a year, and
arises from the fact that nature, in her products, is not
always equally prolific; and the natives migrate from
point to point, for the means of sustenance.

AS MARINERS.

The Marshall islanders are the best and most
skillful navigators in the Pacific.   Their voyages,
sometimes of many months' duration, in great canoes,
sailing with outriggers to windward, well provisioned
and depending on the skies for fresh water, help to
show how the Pacific was colonized. They have a
sort of chart, *mede*, of small sticks tied together,
representing the position of islands and the direction
of the winds and currents. They have also wonderful
weapons, the blades of which are edged with sharks'
teeth, and a defensive armor of braided *sennit*, also
peculiar to the islands.   In hollowing out their canoes
they use a large adze, made from the *Tradacue gigas*,
formerly used in the Carolines, probably by the older
builder race.

LANGUAGES OF MICRONESIA.

The languages of Micronesia, though gramat-
ically alike, differ widely in their vocabularies.   The
religious myths are identifiable with the Polynesian;
but a belief in the gods proper is overshadowed
by a general deification of ancestors, who are supposed from time to time to occupy certain blocks of

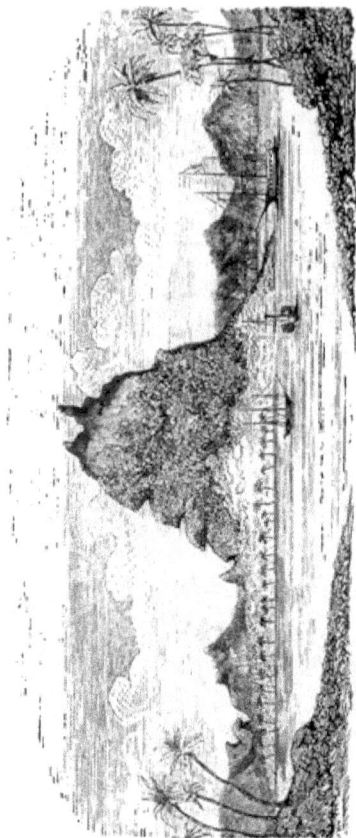

BORA BORA—SOCIETY GROUP.

coral, set up near the family dwelling, and surrounded by circles of smaller ones. These stones are annointed with oil and worshiped with prayer and offerings, and are also used for purposes of divining, in which, and in various omens, there is a general belief. In the Marshall group, in place of these stones, certain palm-trees are similarly enclosed. The spirits, also, are believed to inhabit the forms of certain birds or fishes, which are *tabu*, as food to the family; but they will help to catch these for others. All this closely recalls the *Kauwari*, or the ancestral images of New Guinea.

FLORA AND FAUNA.

The flora of the Gilbert and Marshall groups is of the usual oceanic character, with close Indo-Malay affinities. It is much poorer than that of the Carolines, with their Mollucca and Philippine elements, and this again is surpassed by that of the Ladrones. In the Gilberts, the scattered woods of the cocoanut and *pandanus* have little undergrowth, while the South Marshalls being within the belt of constant precipitation, have a dense growth of low trees and shrubs, with here and there a tropical luxuriance unusual in atolls.

The pandanus grows wild and profusely, and is of exceptional importance, being the chief staple food, so that the cocoanut, which however flourishes chiefly in the Gilberts, is used mainly to produce oil for exportation. The bread-fruit grows chiefly in the South Marshalls. The taro *arum cordifolium* and others is cultivated laboriously, deep trenches being cut in the solid rock for its cultivation. Various veg-

4*

etables grow on soil imported for the purpose. Marine plants are rare.

The fauna, like the flora, becomes poorer eastward, birds being more numerous on the high islands than on the atolls, where the few are chiefly aquatic. On Bonabe, or Ponape, out of twenty-nine species eleven are sea-birds, and of the remaining eighteen, eleven are peculiar to the islands. From the Pelew Islands fifty-six species are recorded (twelve peculiar), and from the neighboring Makenzie group twenty (six peculiar). Yet curiously no species is recorded to those two groups, and peculiar to them. The common fowl is found everywhere, wild or tame, and in some places is kept for its feathers only. The rat and *paunopes* are the only indigenous land mammals. The Indian crocodile is found as far west as the Pelews. There are five or six species of lizards, including a *gecka* and *abliphereos*. Insects are numerous, but of few kinds. Scorpions and centipedes are common, but are said to be harmless.

The houses of the Gilberts and Marshalls (much less elaborate than those in the Carolines) consist merely of a thatched roof, resting on posts, or blocks of coral, about three feet high, with floors at that level, which are reached from an opening in the center. On these the principal people sleep, also serving as a store-house, inaccessible to rats, which infest all the islands.

(Findlay's N. Pacific; Hale's Eth. and Phi. of Wilke's U. S. Ex. Exped.; Menicke's Die Inseln des Stellen Oceans; Proc. Zool. Soc., 1872, 1877, Ency. Brit., vol. 16.)

MICRONESIA.

The Islands of Micronesia lie along the Equator and a little west of the meridian on which the world's day begins. The Micronesian Christians have finished the Sabbath worship, and fallen asleep under the shelter of their thatched cottages beneath the cocoanut trees, before Christians in America have begun the services of the day.

Micronesia is a subdivision of Polynesia, the generic name for the myriad islands scattered over the broad Pacific Ocean. It is composed of four groups—the Gilbert or Kingsmill Islands, which lie on both sides of the Equator and a little beyond the 180th meridian ; the Marshall or Mulgrave Islands, subdivided into the Radac or Ralack Chains ; and the Caroline and Ladrone Islands. The three former groups only are missionary ground, as the Ladrone Islands are a Spanish penal colony, and the native race is extinct.

The Islands of Micronesia are in the great coral belt ; the Gilbert and Marshall groups being exclusively of coral formation, and lie in the Caroline archipelago, which stretches over the sea a distance of two thousand miles from east to west. Many of the atolls or coral islands enclose lagoons from ten to fifteen miles broad, and from twenty to thirty miles long.

The climate of Micronesia is a never-ending summer, never as hot as the hottest summer days of America, and never cold enough to cause chilliness. The greatest range of the thermometer experienced during a residence of several years on Ponape, one of the Caroline group, was thirteen degrees—from 74 deg. to 87 deg. in the shade. On some of the islands the rainfall is excessive; on others, but moderate.

The Islands of Polynesia are inhabited by two races of people—brown and black. The brown are found on the Sandwich Islands, the Marquesas, the Society and the Samoan groups, the Hervey and New

Zealand.   To this race belong the inhabitants of Micronesia.   The Melanesians—found on the Fiji Islands, New Caledonia, the New Hebrides, the Loyalty and Solomon groups, New Britain and New Guinea—are akin to the African, having the woolly hair and physiognomy of the negro races.   They are lower down in the scale of civilization than their brown neighbors, being, as a rule, cannibals—fierce, warlike, treacherous and intractable.   It was among these people that John Williams, Bishop Pattison, the Gordons and other missionaries lost their lives.   But, degraded as they are, the entire history of Christian missions can show no greater transformation than has taken place in the Fiji Islands, as the result of English Wesleyan missions.

The islands inhabited by the black Polynesians enter like a wedge among those inhabited by the brown race, the apex being the Fiji Islands.   The accepted theory, until recently, was that the brown Polynesians belong to the Malay race.   Later investigations by Judge Fornander, of the Hawaiian Islands, and certain German scholars, render it probable that they may be a branch of the Caucasian race.   It is thought that by means of their languages, traditions and mythologies, the Polynesians can be traced back from their present abode, step by step, through the island groups of the Pacific and Indian Oceans, to the Indian Peninsula, and onward to the centre table-lands of Asia, whence the Caucasian races, in the beginning of history, emigrated westward and southward.   In those groups in which the different islands are near enough to allow of communication, even though comparatively infrequent, there is usually a common language ; where widely separated, different languages have been developed. Most of the various dialects abound in vowel sounds, two consonants rarely coming together in the middle of a word, and all words ending in vowels.

Religious beliefs and observances varied with different groups, yet had certain characteristics in common. The people were not idolaters : they believed in the existence of spiritual beings, whose power they feared, and whose anger they sought in many ways to avert. But we never found any conception of a *supreme* Deity, or a belief in one spirit surpassing all others in power. They believed that the spirit of man survived his death, and lived on in one of two places or states, one more desirable than the other, but with no difference based on clearly defined desert or moral character. On some of the islands there was a regular priesthood, with rites of worship; on others, little more than certain superstitious observances. They prayed to spirits, and offered gifts and oblations. Their traditions and mythologies were usually only a confused jumble, and their religious beliefs seemed to have little influence on their character.

(Rev. Robert W. Logan, Congregational Missionary to Micronesia.)

PHŒNIX, ELLIS, UNION AND KERMADEC GROUPS.

To the east of the Marshalls, between latitudes 2 deg. and 5 deg. S., and longitudes 170 deg. and 176 deg. W., are the Phœnix, Swallow, Gardner, Enderberg, Sidney, Hull, Birui, Arthur, Wilkes, and some smaller islets and atolls, sometimes known as the Phoenix Group.

Like many, they are now of no special import-

ance, in size or products.   They but await the occupation and development of the more civilized races, to
render them of great value.

These islands, atolls, and islets although something over a hundred in number, are so similar in
nearly every respect to the Marshall and Gilbert
groups, that a description would be but a repetition of
nearly all that has been written of the latter islands.

Another small group that might be placed under
this head, if we except climatic and geographic differences are the Kermadec islands.   Lying to the north
and east of New Zealand, between latitudes 30 deg.
and 33 deg. S., and about 177 deg. and 179 degrees
W. longitude, might prove of great value, by occupation.   Sunday, Macauley and Curtis islands are the
principal in this little cluster.

NAVIGATOR (SAMOA) ISLANDS.

Samoa, the native name of the Navigator group,
comprises ten islands that are inhabited, or of any
note, with some smaller islets, of no present interest.

Savaii, Opolu, Tutuila, Mauono, Apolima, Mauna,
Olosenga and Of'u are the principal, for a better idea
and description of which I have had to refer to Mr. Reed
of the Australian Customs, and the United States
Exploring Expedition, under Commodore Wilkes, who
surveyed them in 1839.

PHYSICAL FEATURES.

They are located between latitudes 12 deg. 53
min. and 15 deg. 57 min. south, and between longitudes 168 deg. 6 min. and 178 deg. 21 min. west,

with an area I have set down at 1,650 square miles (although some authorities do not allow over 1,100 to 1,200 square miles), with a total population of 35,000. The modern name of the group was given to them by the French navigator, Bougainville, who visited them in 1768. They were visited, also, in after years by the ill-fated la Perouse, in 1787, who had a battle with the natives, losing a good many men in the conflict.

The islands are evidently of volcanic origin, but no traces of active eruptions are found at present. In 1867 a curious submarine convulsion took place in the strait between the islands of Mauna and Olosenga. The eruption lasted for about two weeks, ejecting mud, sand and water in large volumes to a great height. After the convulsion, which in no way disturbed the adjoining islands, the sea flowed peacefully over the volcanoes' watery tomb. Soundings taken at the time showed no apparent variation from the usual depth of water in the strait.

The people are among the straight-haired races of the South Sea. With a fertile soil, blessed with an abundant rainfall, and schools and churches in every village, the group may safely be classed among the garden-spots of the Pacific.

Savaii is the most western island of the Samoan group, and is also the largest, being forty miles in length and twenty in breadth. It is not, however, as populous or as important as some of the others. It differs from any of the others in appearance, for its shore is low, and the ascent thence to the center is gradual, except where the cones of a few extinct craters are seen. In the middle of the island a peak rises, which is almost continually enveloped in the

clouds, and is the highest land in the group. On
account of these clouds angles could not be taken
for determining its height accurately, but it certainly
exceeds 4,000 feet.

Another marked difference between Savaii and
the other larger islands is the want of any permanent
streams, a circumstance which may be explained, notwithstanding the frequency of rains, by the porous
nature of the rock (vesicular lava), of which it is
chiefly composed. Water, however, gushes out near
the shore in copious springs, and when heavy and
continued rains have occurred, streams are formed in
the ravines, but these soon disappear after the rains
have ceased.

The coral reef attached to the island is interrupted to the south and west, where the surf beats
full upon the rocky shore. There are in consequence
but few places where boats can land, and only one
harbor for ships, that of Mataatua; even this is
unsafe from November to February, when the north-
westerly gales prevail. The soil is fertile, and was
composed in every part of the island that was visited,
of decomposed volcanic rock and vegetable mold.
Upolu is ten miles to the eastward of Savaii, and is
next in size. It is about forty miles long and thirteen
broad. It has a main ridge extending east and west,
broken here and there into sharp peaks and hum-
mocks. From this main ridge a number of smaller
ridges and broad gradual slopes run down to a low
shore encircled by a coral reef, interrupted here and
there by channels which form the entrances to safe and
convenient anchorages for small vessels. At Apia
the reef extends across a good-sized bay, and forms
a safe and commodious harbor for large ships, with

an entrance through a deep and clear channel formed by a break in the reef.

Between Savaii and Opulu are two small islands: at the southeast end of Tutuila there is the small island of Aunu'u, and sixty miles to the east of this Maun'a. Of these islands the Rev. Mr. Powell, of the London Missionary Society, says:

"The first island that come, in sight of voyagers arriving from the eastward is Ta'u, the largest of the three islands that constitute the group, which the natives call Manu'a. It is about six miles long, four and a half broad, and sixteen in circumference, and contains one hundred square miles. [This is an evident mathematical mistake of Mr. Powell, as under his description, taking length, breadth or circumference, the island could not contain more than twenty-five to twenty-seven square miles.] About six miles west of Ta'u is the island of Olosenga. This is a very rocky island, three miles long, 500 yards wide, and about 1,500 feet high."

Savaii and Opolu contain the largest extent of flat land; fully two-thirds of their area (about 500,000 acres) are fit for cultivation. The whole group is of volcanic origin; craters of extinct volcanoes are seen at various points. Some of the small islands of the group are composed of a single large crater rising abruptly from the sea. The soil on all the islands is exceedingly rich, and is everywhere covered with dense vegetation from the water's edge up to the tops of the mountains. The high mountain ridges extending through the middle of the larger islands attract the passing clouds, which furnish a copious and never-failing supply of moisture, and feed the numerous streams of beautiful, clear water that abound in every direction.

The climate is mild and agreeable ; the temperature generally ranges between 70 deg. and 80 deg., but the heat is greatly subdued by the breezes that are constantly blowing. Mr. Williams, the British Consul, kept a meteorological register for the Board of Trade from 1860 to 1865, from which I made an abstract of the mean recorded temperature in every month in the year 1864. The southeast trade-winds blow steadily from April to October, being strongest in June and July. From November to March westerly winds frequently blow, but not for any length of time together. A strong gale may generally be looked for some time in January, but frequently an entire year will pass without a severe storm. February, as a rule, is fine, with variable winds. March is usually the worst and most boisterous month of the year, the winds being still variable, and gales occurring from north to northwest. Copious rains fall from the beginning of December to March. June and July are the coolest, and September and October the hottest months; although it will be seen, from the abstract above referred to, that there is very little variation of the temperature throughout the year. Hence the growth of vegetation goes on without check all the year around. Cotton and Indian corn will yield three crops a year. I saw some of the latter gathered in January, which had been sown at the beginning of October. Thus it was planted and the crop gathered within four months. The taro also comes to maturity in four months, and is planted continuously all the year round. When the natives take up the taro, they cut off the top, make a hole in the ground with a stick, into which the top is thrust, without the ground being dug over or in any way prepared. A short time after it is planted, they clean the ground and mulch between the plants with grass and leaves to keep down the seeds. Bananas yield ripe fruit in nine months after planting, some of the introduced varieties coming to maturity in six months. This fruit attains a great size, especially the indigenous varieties, some of which I measured and found to be eight inches long and nine inches in circumference.

PRODUCTS.

The following are the principal productions of the group: Cocoanuts, cotton, native chestnuts, candlenuts, bananas, plantains, oranges, lemons, limes, citrons, pineapples, mangoes, guavas, Malay apples, rose apples, custard apples, pawpaws, tamarinds, bread-fruit, sweet casava, indigo, coffee, Indian corn, tobacco, chile and medicinal plants, several trees with very fragrant blossoms that might be used in the preparation of scents, some that exude aromatic gum, and others that furnish very handsome and durable wood, suitable for cabinet ware and furniture.

INHABITANTS.

The Samoan natives are a fine, tall, handsome race, of a light brown color. They are docile, truthful and hospitable, and are very lively and vivacious. In conversation among themselves, and in their intercourse with foreigners, they are exceedingly courteous and polite. They have different styles of salutation, corresponding with the social rank of the persons addressed; for instance, in addressing the chiefs or dis-

tinguished strangers, they use the expression *Lau-Afio,*
or "Your Majesty;" in speaking to chiefs of lower
rank, they address them, *Lau-Susu,* as we would use
the words "Your lordship;" to chiefs of lower degree
than those who are thus addressed, the term *Ala-Ala*
is used, and to the common people the salutation is
Omai, Sau, simply meaning "You have arrived," or
"You are here."

The men only, tattoo, and not on their faces, as
the New Zealanders do, but on their bodies from the
waist to the knee, entirely black for the most part,
except where relieved here and there by graceful
stripes and patterns. At a short distance this tattooing gives them the appearance of having on black knee
breeches. The clothing of both sexes is a piece of
calico or native cloth wound around the waist and
reaching to the knees. Some of the women wear a
couple of colored cotton handkerchiefs, in the shape of
a narrow poncho, over their breasts and shoulders,
and hanging loosely down to below the waist. When
in the bush, or working on their taro plantations, or
when fishing, they wear a kilt of the long, handsome
leaves of the Ti (*Dracæna terminalis*). They have a
kind of fine mat plaited from thin strips of the leaves
of a plant called *Lau-ie.* These mats are only used on
important occasions, and they esteem them more highly
than any European commodity. Some of these mats
are quite celebrated, having names that are known all
over the group; the older they are the more they are
valued. The oldest one known is called *Moe'efui-fui,*
meaning "the mat that slept among the creepers."
This name was given to it from the circumstance of its
having been hidden away among the creeping kind of
convolvolus that grows along the shores; it is known

to be over two hundred years old, as the names of its
different owners during that time can be traced down.
The best mats are made at Manu'a. They are the
most coveted property a native can possess, no labor
or enterprise being considered too great to secure
them. Both men and women spend a great deal of
of time in dressing their hair, and frequently apply
lime to it, which is laid on in a liquid state about the
consistency of cream, and has the effect of turning the
hair to a reddish hue. Both men and women frequently wear flowers in their hair—generally a single
blossom of the beautiful scarlet Hybiscus, which is always found growing near their houses. Nature has
supplied them so bountifully with food, in the shape of
the cocoanut, bread-fruit, banana, native chestnuts, and
other wild fruits, and the taro yields so abundant a
crop with so little cultivation, that they have no necessity to exert themselves much, and they are, therefore,
little inclined to industry, and probably will never be
induced to undertake steady labor of any kind. Their
houses are neat, substantial structures, generally circular in shape, with high, pitched, conical roofs, supported in the centre by two or three stout posts, and
open all around, but fitted with narrow mats made of
cocoanut leaves, which are strung together like Venetian blinds, and can be let down in stormy weather.

The Samoans are very expert in the management
of their canoes, of which they have five different kinds
—the *Alia*, or large double canoe, some of which are
capable of carrying two hundred men ; the *Tau-mau-
lua*, from thirty to fifty feet long—these were first made
about thirty years ago, and are fashioned after the
model of our whale-boats ; the *Va-lao*, or fishing canoes, with out-riggers—a beautiful craft, and very fast ;

then there is the *Loatau*, out-rigger, dug-out canoe,
capable of carrying five or six people; and, lastly, the
Paopao, a small dug-out canoe for one person.

RELIGION.

The natives are all professed Christians. Christianity was first introduced into Samoa in August, 1830,
by the Rev. J. Williams, who landed a number of native teachers from Tahiti. A few years afterwards
(about 1835) five English missionaries, belonging to
the London Missionary Society, landed on the islands,
and from that time to the present several Congregational missionaries have been constantly resident on
the group. In addition to these, there is a Roman
Catholic Bishop resident at Apia, and a number of
Catholic priests in various parts of the islands. The
natives, for many years past, have annually contributed
considerable sums towards the support of the mission
establishments.

These islands, in their varied productiveness and
their great capabilities for immense agricultural returns, if put under a proper system of cultivation, with
the habits and manners of the inhabitants, are a fair
type of the most of the groups of the Pacific. At the
present day they are living proofs of the incalculable
benefits that may arise from the gradual American
protectorate, with its modern methods and appliances,
spreading over these regions.

METEOROLOGICAL.

Extract from the Meteorological Register kept at
the British Consulate at Apia, in the Navigator

Islands, which may be accepted as about the temperature of all of the tropical islands of the Pacific.

MONTH.	LOWEST AND HIGHEST TEMPERATURE DURING THE MONTH.				HIGHEST RECORDED TEMPERATURE DURING THE MONTH.
	6 A. M.		4 P. M.		
	Lowest	Highest	Lowest	Highest	
January	70	75	76	82	85—at 8 A. M.
February	71	79	77	84	85 " 10 "
March	70	81	74	85	86 " 8 "
April	70	76	74	88	88 " 4 P. M.
May	65	82	78	85	85 " 4 "
June	65	74	78	83	83 " 4 "
July	61	74	79	82	82 " 4 "
August	59	77	78	84	84 " 4 "
September	67	78	81	83	86 " 8 A. M.
October	61	79	82	84	86 " 8 "
November	73	76	78	79	84 " 8 "
December	71	78	82	86	86 " 4 P. M.

BANKS ISLANDS.

North of the New Hebrides we come to the Banks group, named after Sir Joseph Banks, scientist and naturalist, who accompanied Captain Cook in his voyage to the Society Islands in 1768.

"Vanua Lava, the largest of the group, is fifteen miles in length north and south, and is a remarkable looking island, with several high, rounded mountains, the highest, to the northwest, being some 2,800 feet above the sea. In the Suutamiti Mountain are several hot springs, always steaming, whilst a stream impregnated with sulphur runs down to the sea on the northwest coast, and a similar one falls into Port Patterson on the Eastern side. There are two waterfalls on the western side—one single and the other double. The population of Vanua Lava is about 1,500; the natives were quiet and friendly."

This island, with Santa Maria, Mota, Valua, Arau and Ureparapara, with some smaller islets dotting the sea, make up the group. The inhabitants are quite friendly with strangers, although very quarrelsome among themselves. This may be attributed to their desire to trade for the curiosities (to them) in the possession of the whites. Anything, from a small piece of hoop-iron to a chopping-axe, is eagerly bartered for.

The weapons of the natives are bows and poisoned arrows, war-clubs and spears, which they handle with the greatest dexterity. The products are fruit, sugar-cane, taro, potatoes and yams.

SANTA CRUZ ISLANDS.

Still pursuing our northerly course, we arrive at Santa Cruz Islands, composed of seven larger ones. Volcano, Vuerta, Santa Cruz, Edgecombe, Ouvry and Lord Howe, with several smaller ones; Vanikoro is made interesting in a historical way, from having been the scene of the wreck of the two vessels under command of Admiral de la Perouse, the great French voyager. This occurred in 1788, and remained an uncertainty for many years, causing much uneasiness in his native land, and, in fact, all over the civilized world. In 1826 the chains, anchors, cannon and some of the heavier imperishable portions of his vessels were discovered at this island and taken to France, in memory of Perouse.

Of Santa Cruz, Captain Tilly says: " It is about

EARLY DAYS IN AUSTRALIA.

.

fifteen or sixteen miles in length, with fringe reefs along the shore, but apparently no off-lying dangers. The north point near the center of the island was found to be in latitude 10 deg. 40 min. south, and longitude 166 deg. 3 min. The high land extends close out on its northeast side, but towards the northwest the hills slope at some distance from the extremes, leaving a considerable extent of low land near the coast. The island is well wooded and watered, the streams in some places running through the villages into the sea.

"The natives are a fine athletic race, and came off readily to the ship, bringing pigs, bread-fruit and yams. Mats, in the manufacture of which great skill is displayed, are also offered for sale. The appearance of the canoes, houses, etc., evinces great ingenuity. Canoes with outriggers, and mostly lime-washed, have a neat appearance; they have also large sea-going double canoes. The villages are large, and houses surrounded by stone fences. On the north side, the villages are close to the sea, with from 300 to 400 inhabitants each. The natives are apparently merry and good-natured, but are not to be trusted, for without any known reason they attacked the boat of the Bishop, on leaving the village of the northwest extremity of the island, and nearly succeeded in cutting it off. Three of the crew were wounded with arrows, and of these two died from the effects of their wounds. Their bows are formidable looking weapons, being seven feet in length, with arrows in proportion."

SOLOMON ISLANDS.

North by west from Santa Cruz is the Solomon Archipelago, so named by Mendana, the discoverer, in 1568, supposing the islands to contain all the wealth and riches like unto that of the ancient king. They were re-discovered by Phillip Carteret, in 1767. The group is very extensive, ranging many hundreds of miles northwest and southeast, although but eight or ten are well enough known to afford data for a description. The principal are Malayta, Ysabel, Guadalcanar, Bougainville, San Christoval and Choiseul. They are of large size, some being fully 100 miles long by twenty or thirty miles wide, with lofty ranges of mountains sloping gradually to the sea, well watered and covered with trees and ferns, with here and there beautiful valleys, and streams of water meandering through them to the sea.

The inhabitants are active and energetic, and are great mariners, their canoes being well built, and handled with consummate skill. Some of their war-canoes are fully eighty feet long, with beam of five feet, and carry sixty men. They are very skillful in carving, while many of their weapons and industrial implements are inlaid with the mother-of-pearl shell.

CHAPTER V.

———

———

Deep in the wave is a coral grove,
Where the purple mullet and gold fish rove.

JAMES GATES PERCIVAL.

CAROLINE ISLANDS.

THE Caroline group, extending almost from the equator to 12 deg. north latitude, and ranging from 135 deg. to 177 deg. east longitude, comprises over 500 islands. Dotting the great Pacific Sea with lands of indescribable fertility and fabulous commercial possibilities, they are almost beyond the description of tongue or pen. If anything were needed to substantiate the grandeur and extent of some of the islands and atolls of the Pacific, the following description would alone suffice.

THE GREAT ATOLL OF IIOGOLEU.

Lying at the eastern end of the great Caroline group, it surrounds and contains within its limits a principality. If one could imagine a strip of land

five to eight miles wide, varied in its topography by
mountain, hill and valley, traversing the ocean for
nearly 300 miles, in almost the form of a circle, and
this strip covered with the most beautiful tropical
foliage, of fruit and other valuable trees, some idea
of the outward form of Hogoleu might be obtained.
Enclosed in this great circle of land lies the lagoon,
with four greater and twenty smaller islands dotting
the surface, on whose broad expanse of waters the
combined navies of the world might ride at safe
and roomy anchorage. With three main outlets to
the ocean, whose width and depth render them perfectly safe for the passage of the greatest ships,
the lagoon forms an inland harbor unequalled in any
other part of the world. The islands in the lake,
some of which are thirty to forty miles in circumference, are covered with valuable timber, and abound
in all the tropical fruits, of the cocoanut, citron,
bread-fruit, oranges, bananas and mangoes, with trees
of the sago and date palm, and timber of the toa,
tomano, prima vera, and great quantities of sandalwood. Fine streams of fresh water flow through the
valleys, while to add to the gorgeous beauty of the
scene, birds with the most beautiful and valuable
plumage give life and animation to the forests and
glades. Here, too, the beche-de-mer, the tortoise
and turtle find their favorite breeding-grounds, in the
water and along the shores. The great lake teems
with fish of nearly all the species found in the South
Sea, many of whose brilliant hues and colors are
only equalled by the pearl shell that line the bed
of the lagoon. The latter is found here in great
abundance, of the largest size and finest quality,
covering the bottom of the lake wherever it can be

seen, and of course in just as great if not greater abundance in the depths not reached by the eye.

Our limited stay at Hogoleu hardly gave me time to form a just opinion of the character and manners of the natives, for which I have been forced to rely upon the valuable experiences and writings of others.

INHABITANTS.

"In judging of the character of the Caroline Islanders, one must remember that there are always two sides to a question; and in connection with this matter, I may refer to a fact which I regard as very significant. All Englishmen are familiar with the story of the wreck of the *Antelope* at the Pelew Islands in 1793, and of the Prince Lee Boo, who accompanied Captain Wilson to England. These same Pelew Islanders, who at that time treated the shipwrecked Englishmen with such generous hospitality for a period of four months, seeking no return for the same, are now regarded as piratical miscreants of the most atrocious type—and not without reason, for they have got into a bad habit of going out to sea in their fast-sailing proas, and attacking, off the coasts of their islands, such vessels as may be becalmed or entangled among the shoals; in which nefarious practice they have, on several occasions, so far succeeded as to have plundered the vessels and massacred their crews. This change of behavior is easily accounted for. In some cases it has arisen from ill treatment which they have experienced at the hands of strangers, but in most cases it has been the result of evil example by a set of scoundrels who disgrace humanity, and are to be found strolling about these seas, making themselves at home

among the simple-minded barbarians, and instructing
them in every vice and villainy.

"No one knows with any certainty how many inhabitants are on Hogoleu; some say 15,000, some
20,000; but there are very many. They are armed
with good swords with hilts of brass, daggers, spears
pointed with iron, bows of great strength, arrows
headed with iron, and slings out of which they fling
round stones with great certainty and with the force
of a shot. The iron weapons they have purchased
from traders of Manilla and elsewhere. They have
many combats with crews of ships, and display great
courage. No white men have ever lived among them,
to anyone's knowledge, though I have heard there is
one living there now, established by one Captain
Hayes. Many men have been on shore and have
been treated with hospitality. From what I have seen
of them, they are a people I would have no fear of,
although they have an ugly habit of attacking ships
upon small grounds of offense. In 1870 they tried to
board the *Vesta*, but the German captain, although he
lost his anchor and chain by having to slip it, was more
than a match for them. He fired upon them with
scrap-iron and killed a great many. Of course, he was
not to blame; but these unfortunate misunderstandings
tend very much to perpetuate ill feeling.

"That the first Europeans who can succeed in
establishing a permanent agency upon Hogoleu will
make their fortunes in a very short period, is an un-
questionable fact. This island presents to the commercial adventurer such an opportunity as is scarcely
to be found elsewhere in the world—not alone from
the valuable products of the land itself, but from the
possession of so magnificent a harbor for shipping,

whence could be extended the ramifications of a trade on a large scale throughout the whole great Caroline Archipelago. That there is any risk in the attempt, I do not for a moment believe. All that is required is for one determined man, acquainted with the Caroline tongue, to secure, by acceptable presents, the protection of a chief, to marry into his family (as he would be required to do), and after a few months' diplomacy he might have it all his own way, so far as driving a trade for his owners was concerned."

PELEW ISLANDS—ATOLLS.

The Pelew Islands referred to form the extreme western end of the Caroline group, and were discovered by Drake in 1579; the main Carolines having been visited by Alonzo de Saavedra, as early as 1528, although the discovery of the group has been ascribed by some writers to Lopez Villa Lobas, in 1543, which is an evident mistake.

These Atolls, or horse-shoe islands (sucl. as I have described Hogoleu), are an important feature in the geological formation of the Pacific Isles, and are to be found in nearly every group, as well as scattered over the great waste of waters of the South Sea, sometimes isolated and alone, at others in groups and chains, having the appearance of the last outposts of a sunken continent. Darwin, Humboldt and others account for their singular shape and formation by assuming that at one time they were portions of the mainland or continents, or islands, and that their centers, which at former periods were hilly and mountainous, gradually sank and disappeared; the coral insect building the fringe or edge on the sunken lands in the form we now see them. They vary somewhat in size and form, and may be found from but a mile or so in diameter to hundreds of miles in circumference.

The inland lakes are nearly always safe harboring for vessels sailing and trading in these seas. Generally speaking, there are from one to four openings or passage-ways from the sea to the lagoons, through which the tide ebbs and flows. These channels vary from fifty to several hundred yards in width, and carry deep navigable water. In the storms and gales that sometimes prevail in these regions, an atoll might be truly termed the sailors' snug harbor.

CORAL REEFS.

A wide platform of rock, covered with the sea, except at low tide, borders most of the high islands of the Pacific. It is a vast accumulation of coral, based upon the bottom in the shallow waters of the shores. This bank or table of coral rock is of varying width, from a few hundred feet to a mile or more; and although the surface is usually nearly flat, it is often intersected by irregular boat channels, or occasionally incloses large bays, affording harbor protection to scores of ships. In very many instances it stands at a distance from the shores, like an artificial mole, leaving a wide and deep channel between it and the land, and within this channel are other coral reefs, some in scattered patches and others attached close to the shore. The inner reef in these cases is distinguished as the *fringed* reef, and the outer as the *barrier* reef. The sea rolls in heavy surges against the outer margin of the barrier; but the still waters of a lake prevail within, affording safe navigation for the tottling canoe, sometimes through the whole circuit of an island; and not unfrequently ships may pass, as by an internal canal, from harbor to harbor around the island.

The reef is covered by the sea at high tide, yet the smoother waters indicate its extent and a line of breakers its outline. Occasionally a green islet rises from the reef, and in some instances a grove of palms stretches along the barrier for miles, where the action of the sea has raised the coral structure above the waves.

Coral islands resemble the reefs just described, except that a lake or lagoon is encircled instead of a mountainous island. A narrow rim of coral reef, generally but a few hundred yards wide, stretches around the inclosed waters. In some parts it is so low that the waves are still dashing over it into the lagoon, and in others it is verdant with the rich foliage of the tropics. The coral-made land when highest is seldom over eight or ten feet in height.

When first seen from the deck of a vessel, only a series of dark points are descried just above the horizon. Shortly after, the points enlarge into the plumed tops of the cocoanut trees, and a line of green, interrupted at intervals, is traced along the water's surface. Approaching still nearer, the lake and its belt of verdure are spread out before the eye, and a scene of more interest can scarcely be imagined. The surf beating loud and heavy along the margin of the reef, presents a strange contrast to the prospect beyond—the white coral beach, the massy foliage of the grove, and the embosomed lake with its tiny islets.
* * Very erroneous ideas prevail respecting the appearance of a bed or area of growing corals. The submerged reef is often thought of as an extended

mass of coral, alive uniformly over its upper surface, and by this living growth gradually enlarging upward; and such preconceived views when ascertained to be erroneous by observation, have sometimes led to skepticism with regard to the zoophyte origin of the reef rock. Nothing is wider from the truth, and this must have been inferred from the description already given. Another glance at the coral plantation should be taken by the reader, before proceeding with the explanations which follow.

Coral plantation and coral field are more appropriate appellations than coral garden, and convey a juster impression of the surface of a growing reef. Like a spot of wild land, covered in some parts with varied shrubbery, in other parts bearing only occasional tufts of vegetation over barren plains of sand, here a clump of saplings, and there a carpet of variously colored flowers—such is the coral plantation.

Numerous kinds of zoophytes grow scattered over the surface, like the vegetation of the land. There are large areas that bear nothing, and others that are thickly overgrown. There is no green sward to the landscape, and here the comparison fails. Sand and fragments fill up the bare intervals between the flowering tufts, or where the zoophytes are crowded; there are deep holes among the stony stems and folia, that seem as if formed among the aggregated roots of the living corals. * * *

These fields of growing coral spread over submarine lands, such as the shores of islands and continents, where the depth is not greater than their habits require—just as vegetation extends itself through regions that are congenial. The germ or ovule, which, when first produced, swims free, finds afterwards a point of rock or dead coral to plant itself upon, and thence springs the tree or some other form of coral growth.

ANALOGY TO VEGETATION.

The analogy to vegetation does not stop here. It is well known that the debris of the forest, decaying leaves and stems, and animal remains, add to the soil, and that accumulations of this kind are ceaselessly in progress ; that by this means, in the luxuriant swamp, deep beds of peaty earth are formed. So it is in the coral mead. Accumulations of fragments and sand from the coral zoophytes, and of shells and other relics of organic life, are in constant progress, and thus a bed of coral debris is formed and compacted.

There is this difference—that a large part of the vegetable material consists of elements which escape as gases on decomposition ; whereas, coral is itself an enduring rock material, undergoing no essential change except the mechanical one of comminution, the animal portion is but a mere fraction of the whole zoophyte.

In these few hints we have the whole theory of reef making ; not a speculative opinion, but a legitimate deduction from a few simple facts, and bearing close analogy on land. The coral debris and shells fill up the intervals between the coral patches and the cavities among the living tufts, and in this manner produce the reef deposit, which is consolidated by the filtrating sea-water, having more or less lime in solution.

(Notes from U. S. Ex. Expedition in 1838, '39, '40, '41 and '42; James D. Dana, A. M., Geol. of Ex.)

AUSTRALIA.

The leviathan of the island groups of the world,
Australia (literally South Asia), lies between latitude
10 deg. 43 min. and 39 deg. 9 min. south, and longitude 113 deg. 15 min. and 153 deg. east, comprising
within its vast limits three million square miles. It
has a sea-coast of over eight thousand miles, along
the line of which eighty-two small islands are located.
Australia was discovered about 1606 by the Dutch,
who were the first to locate it and chronicle its existence in modern times. It was first named by them
New Holland, a name retained for many years.

From the sea this great island-continent presents
an uninviting appearance, giving one the impression
that the crags and mountains fringing the shores enclose a sterile waste within. Probably no country in
the world has received more attention from men of
science and explorers than Australia, and that, too,
with less beneficial results, as the great mountain ranges and barren wastes of the interior are to-day as an
unknown land.

One of the greatest detriments to its rapid progress in peopling and civilization, was its establish-
ment as a penal colony by Great Britain. This, together with the low order of the native races, some
two hundred thousand in number, who are little above
the animal in the scale of humanity, proved for many
years a great barrier to the peopling of the island with
the better classes. Until 1851 the progress of Australia was under a ban; when Mr. Hargreaves, returning from the gold fields of California, discovered the
precious metal on the island. From this time may be
dated the advancement of that country. The gold

fever drew people from all parts of the world to settle on her shores. Cities and towns rapidly sprang into existence, while the consequent development of great agricultural resources, fed with the thousand millions in gold taken from her mines, placed her at once among the great countries of the world. With the single exception of California, nothing like Australia's progress has occurred in ancient or modern times. The discovery of many valuable mines of copper, coal, tin, lead and silver followed that of gold, and being found in large and paying quantities, add largely to the income of the inhabitants.

PHYSICAL FEATURES.

The mountain ranges on the island are but few in number. The greatest altitude of those already discovered does not exceed seven thousand feet.

There are many ponds and swamps in the interior, with few navigable streams—only in the rainy season. Even then navigation is very uncertain, as the waters of most of the rivers frequently disappear —lost in the sands of the surrounding wastes.

The flora of the island is not varied or extensive, but two species forming the principal forest growth— the eucalyptii and acacia—although more than one hundred varieties of each of these interesting species are found, and in great abundance.

GEOLOGICAL AND GEOGRAPHICAL.

The geological formation is quite an interesting study, partaking of the eruptic, metamorphic, trappean, with the sedimentary sandstones of the tertiary period.

From careful scientific observations, it is found that Australia is slowly rising from the deep—gradually but surely taking its place among the continents of the world. Unlike some of its short-lived neighbors lying to the westward in the Straits of Sudan, whose appearance and disappearance mark but a period in the birth, growth and death of islands, Australia is apparently on a foundation that may last for all time. The population is about two millions, who, when not mining, are principally in the agricultural and grazing interests. The value of exports and imports may be stated at $500,000,000 per annum.

The island, from its immense area, is marked off in several colonial divisions. The principal of these are Northern Australia, or Alexandra's Land, colonized in 1838 ; Western Australia, colonized in 1829 ; south from which is Tasman's Land, surveyed in 1818; Southern Australia, colonized in 1834; Queensland on the northeast, and New South Wales on the southeast, colonized in 1778.

Captain Cook is credited with the discovery of Australia in 1770. Tasman, who discovered New Zealand and Tasmania as early as 1642, could not have failed to notice and locate it in his voyages. Dirk Hartog, a Dutch navigator, is credited also with its discovery, by some authorities, in 1606.

NEW ZEALAND.

This group of three principal and thirteen smaller islands, to the southeast of Australia, between latitudes 34 deg. and 48 deg. south and 161 deg. and 179 deg. east longitudes, comprise in their area 122,582 square miles, a little larger than Great Britain and Ireland. The population is 476,000.

The geological formation is volcanic eruptic, with the sedimentary formations and fossils of the tertiary period. Like Australia, the lands are slowly rising from the sea.

Of minerals, the islands have an abundant supply—coal, copper, iron, lead and manganese being found.

The natural vegetation of New Zealand is wonderful in its luxuriance, many hundreds of species crowding the forests. Nearly all of these are of the evergreen type, and give to the islands an appearance of perpetual spring.

Of the animal kingdom there is but little to be said, as when discovered in 1642, by Tasman, a species of rat and the dog were about the only animals to be found. Those of a more recent date are altogether domestic, the results of importations from other countries by the settlers.

From the long narrow configuration of the islands, the streams, though many in number, are of no great length, breadth or depth.

Mountain ranges cross the islands in many places, but generally speaking are not of great prominence, if we except Mt. Cook, which is supposed to be 14,000 feet high. There are many evidences of volcanic action throughout the group. Tongariro is the only active volcano at the present time.

Of the natives, Mr. Taylor says: "The New Zealanders are decidedly a mixed race—some have wooly hair, others brown or flaxen; some are many shades darker than others. The peculiar features of the Mongol are also very common; the oblique eye, the yellow countenance, the remarkable depression of the space between the eyes so that there is no rise in the nose, seem clearly to indicate that some portion of the race is of Chinese or Japanese descent.

Tasmania, or Van Dieman's Land, just south of Australia, between 40 deg. 40 min. and 43 deg. 38 min. south latitude, and longitude 144 deg. 33 min. and 148 deg.. 28 min. east, a group of some seventeen islands occur; but one of them is of any size or importance at present.

Tasmania was discovered and located by Tasman in 1642, but was re-located and taken possession of by the English in 1803. The island has an area of 22,629 square miles, with a population of 110,000.

The island is of a similar formation to Australia, although the soil is much more fertile, and without any of the desert wastes of the larger island. The mountain ranges are extensive, but not of very great height. The forests are immense, the eucalyptus and acacia, in all their many varieties, growing in the greatest luxuriance.

Of minerals, Tasmania has an abundance—gold, copper, iron and coal mines are worked at a considerable profit.

The climate is temperate; all the fruits, vegetables and cereals are cultivated, forming one of the principal exports of the group.

The natives are of the same type as the aborigines of New Zealand and Australia, and are now nearly extinct.

COFFEE PLANT—JAVA.

CHAPTER VI.

ISLANDS

The place is all awave with trees,
Limes, myrtles, purple-beaded;
Acacias having drunk the lees
Of the night-dew, faint-headed;
And wan, grey olive-woods, which seem
The fittest foliage for a dream.

     E. B. BROWNING (*An Island*).

JAVA.

THE Island of Java, with its 52,000 square miles, peopled by nearly eighteen millions of inhabitants—the "land of fire," the home of the eruptic volcano and earthquake—has long been the subject of interesting study for the historian and scientist.

Here we find, besides innumerable smaller ones, one of the largest volcanic craters in the world, having a circumference around its edge of about twelve miles. In 1772 this crater was in active force, casting its ashes and scoria over a great tract of country. Thousands of inhabitants lost their lives—either caught in their homes by the burning lava, or suffocated by the smoke, ashes and sulphur. The heavens were lit up for hundreds of miles around with a glare only equalled by that of the *aurora borealis*, the surrounding seas liter-

6*

ally covered with the finer particles of pumice and
ashes, while the dust and smoke hung in and darkened
the heavens for days afterwards.

Another eruption took place in 1832, with the loss
of nearly thirty thousand lives, and again in 1883, when
it is supposed one hundred thousand people were destroyed, with a vast waste created over a beautiful and
thriving agricultural country.

GENERAL FEATURES.

The topographical features of the island, its chains
of mountains and plateaus, with the valleys lying between, the latter well watered by meandering rivers,
are nearly all taken advantage of by a skillful, agricultural people. The waters from abundant rainfalls
are treasured in reservoirs on the higher plateaus, and
held in reserve for the drier periods. They are thus
enabled to reap two crops per annum, and place their
plantations in almost continuous bloom. On the cultivated lands, immense quantities of coffee, sugar, rice
and cotton are grown, with all the fruits of the tropics,
as well as the clove, nutmeg and cinnamon, and other
spices.

Included in the flora of the native forests are the
gutta percha, toa tomano, camphor, sandal, satin-wood
and mahogany trees.

The agricultural methods adopted by the natives,
with the use of irrigation, was imparted to them by the
Hindoos and others of the East India countries, who
visited this island in great numbers many years previous to the ninth century.

The inhabitants at present are hospitable and intelligent—partaking of the higher class of Arabs in

character and religion. The Mohammedan belief is
general, having been forced on the Javanese by the
Arabs in the 15th century.

MICHELET ON JAVA.

Of Java, Michelet, the great French writer, says:
It is dowered with fires. Notwithstanding its limited
area, it possesses as many as the entire continent of
America, and all of them more terrible than burning
Etna. And to these we must add its *liquid volcano*,
its vein of somber azure which the Japanese call the
"Black River." This the great Equatorial Current,
which in its northerly course warms the Asiatic seas,
is remarkable for its muddiness, and tastes salter than
human blood.

A hot sea—a torrid sun—volcanic fire—volcanic
life! Not a day passes but a tempest breaks out
among the Blue Mountains, with lightning so vivid
that the eye cannot endure to gaze at it. Torrents of
electric rain intoxicate earth and madden vegetation.
The very forests smoking with wreathed vapors in the
burning sun, seem so many additional volcanoes situated midway on the mountain slopes.

In the loftier regions, they are frequently inaccessible, and sometimes so thickly intertangled, so
dense, so gloomy, that the traveler who penetrates
them must carry torches even at noonday. Nature
without an eye to watch her, celebrates there her
"orgies of vegetation," and creates, as Blum informs
us, her river monsters and colossi.

Stemless rhizanthæ seize on the roots of a tree
and gorge themselves with its pith and vitality. Travelers speak of a species which measures six feet in

circumference. Their splendor, shining in the deep
night of the forest, astonishes, nay, almost terrifies
the spectator. These children of the darkness owe
nothing of their resplendent coloring to the light.
Flourishing low down in the warm vapors, and fat-
tened by the breath of earth, they seem to be its lux-
urious dreams, its strange airy phantasies of desire.

Java has two faces. The southern, wears already
the aspect of Oceania, enjoys a pure air, and is
surrounded by rocks all alive with polypes and
madepores. To the north, however, it is still in
India—India, with all it inherits of unhealthiness ; a
black alluvial soil, fermenting with the deadly travail
of Nature reacting on herself, with the work of combination and decomposition. Its inhabitants have
been compelled to abandon the once opulent town of
Bantam, which is now a mass of ruins. Superb Batavia is one triumphant cemetery. In less than thirty
years—from 1730 to 1752—it swallowed up a million
of human lives ; sixty thousand in a single twelve-month
(1750)! And though it is not so terrible now, its
atmosphere has not been purified to any considerable
extent.

The animals of the primeval world which live
forgotten in its bosom are remarkable, it seems, for
their funeral aspect. In the evening enormous hairy
bats, such as are found nowhere else, flutter to and
fro. By day, and even at noon, the strange flying
dragon, that memorial of a remote epoch, when the
serpent was endowed with wings, does not hesitate to
make its appearance. Numerous black animals exist
which agree in color with the black basalt of the mountains. And black, too, is the tiger, that terrible destroyer,
which as late as 1830, devoured annually 300 lives.

TOPOGRAPHY.

The double mountain chain, which forms the back-bone of Java, is intersected by numerous internal valleys, running in opposite directions, varying the spectacle. This diversity of surface insures a corresponding diversity of vegetation. The soil in the valleys is madreporic, and was once *alive*. At a higher level it has its foundation of granite, loaded with fertile ruins and hot *debris* of the volcanoes. The whole is a vast ascending scale, which from sea to mountain presents six different climates, rising from the marine flora and the flora of the marshes to the Alpine flora. A superb amphitheater, rich and abundant at each gradation, bearing the dominant plants and those transitional forms which lead up from one to the other, and lead so ingeniously that without any lacuna or abrupt leap, we are carried onwards, and vainly endeavor to trace between the six climates any rigorous lines of demarcation.

In the lowlands facing India and the boiling caldron of the ocean, the mangrove absorbs the vapors. But towards Oceania and the region of the thousand isles, the cocoanut tree rises, with its foot in the emerald wave and its crest lightly rocking in the full fresh breeze.

The palm is here of little value. Above its bamboos and resinous trees, Java wears a magnificent girdle, or zone, of forest—a forest wholly composed of teak, the oak of oaks, the finest wood in the world —indestructible teak.   *   *   *

Here every kind of food, and all the provisions of the five worlds superabound. The rice, maize, figs and bananas of Hindostan; the pears of China;

the apples of Japan, flourish in company with the peach, pineapple and orange of Europe—aye, and even with the strawberry, which extends its growth along the banks of the streams.

All this is the innocence of nature. But side by side with it prevails another and more formidable world—that of the higher vegetable energies, the plants of temptation, seductive, yet fatal, which double the pleasures, while shortening the duration of life.

At present they reign throughout the earth, from pole to pole. They make and unmake nations. The least of these terrible spirits has wrought a greater change in the globe than any war. They have implanted in man the volcanic fires; and a soul, a violent spirit which is indefinable, which seems less a human thing than a creature of the planet. They have effected a revolution, which, above all, has changed our idea of time. Tobacco kills the hours and renders them insensible. Coffee shortens them by the stimulus it affords the brain; it converts them into minutes.

Foremost among the sources of intoxication to which care unhappily resorts, we must name alcohol. Eight species of the sugar-cane which thrive in Java abundantly supply this agent of delirium and forcible feebleness. No less abundantly flourishes tobacco, the herb of dreams, which has enshrouded the world in its misty vapors. Fortunately Java also produces immense supplies of its antidote, coffee. It is this which contends against tobacco, and supplies the place of alcohol. The island of Java alone furnishes a fourth of all the coffee drank by man, and a coffee, too, of fine quality, which has been dried sufficiently, without any fear of reducing its weight.

Formerly Java and its neighboring lands were known as spice islands only, and as producing freely violent drugs and medicinal poisons. Frightful stories were circulated of its deadly plants, the juice of which was a mortal venom—of the *Gueva-Upas*, which but to touch was death!

CLIMATE.

He who would see the East in all the fullness of its magical, voluptuous and sinister forces, should explore the great bazaars of Java. There the curious jewels wrought by the cunning Indian hand are exposed to the desires of woman, temptation and the cost of pleasure. There, too, may be seen another seductive agency—the vegetable fury of the burning and scorching plains which is so eagerly sought after: the perfumes of terrible herbs and flowers, as yet unnamed. Marvelous and profound the night, in its sweet repose, after the violent heats of the day! But be cautious in your enjoyment of it; as it grows old it breathes death!

Take note of this: The peculiarity that gives to these bazaars so curious an effect is, that all the thronging crowds are dusky, with dark complexions, and all the animals are black. The contrast is singular in this land of glowing light. The heat seems to have burned up everything, and tinted each object with shadow. The little horses, as they gallop past you, seem but so many flashes of darkness; the buffaloes, slowly arriving, loaded with fruit and flowers—with the most radiant gifts of life—all wear a livery of bluish black.

Beware, at this time of night, not to wander too far, or ramble in the higher grounds, lest you should

encounter the black panther, whose green eyes illu-
mine the obscurity with a terrific glare ! And—who
knows?—the splendid tyrant of the forest, the black
tiger, may have begun his midnight prowl—that formidable phantom which the Malays of Java believe to
be the spirit of Death !

I have quoted thus, at some length, from the writings of Michelet, as the ideas advanced will serve alike
for Sumatra and some of the Mollucca Islands.

Borneo, singularly, is altogether free from the
eruptic, volcanic and earthquake forces. Situated almost directly in the course of the "fire belt," there are
yet no authentic records in the history of Borneo, for
ages past, of any of those fearful outbursts so frequent
in Java and Sumatra.

LITTLE JAVA.

Much more could be written of Java and the islands surrounding it. As almost a part of the greater
island, we might cite Little Java, with nearly four thousand square miles of area, and a population of about
eight hundred thousand people. Separated from
Great Java by a strait hardly two miles in width, its
configuration, climate, inhabitants and products are so
similar that a description would but tire the reader.

COFFEE.

Before leaving Java, it might be well to notice
coffee, the principal and most valuable product of that
island. *Coffea Aribica*, no doubt, derives its name
from Kaffa, a district of Southern Abyssinia, on the
east coast of Africa. The coffee plant is an evergreen,

and was first found growing wild in Arabia. Africa and some portions of South America. It is sometimes cultivated at a height of six thousand feet above the sea-level, but this only in warm countries, as the tree does not thrive in climates where the thermometer falls below 55 deg.

In its wild state the tree grows from ten to thirty feet high, but when cultivated it is pruned down to five or six feet—the yield being greater, while the berry is much easier to harvest. The young plants are usually grown from the seed in nurseries, and when a year old are transplanted to such localities as desired. The tree, in favorable climates, begins to bear fruit at three years, but hardly in paying quantities until the fifth year. From this age the plant bears from two to three crops per annum for twenty years, after which the yield is hardly profitable, when the older trees are replaced with younger plants.

The fruit of the coffee tree greatly resembles the cherry, in size and color, when ripe ; the coffee, as we see it in commerce, being the seeds, of which there are two to each berry. The kernels are extracted, after the fruit is thoroughly dried, by being passed through wooden rollers, which crush and separate the hull from the grains.

The best coffee is Mocha, grown in the province of Yemen, in Arabia ; that from Java taking second place. Brazil is credited with producing something over half of all the coffee consumed in the world, although the quality is not equal to Mocha or Java. It is a little difficult to judge of the brands of coffee offered in the markets nowadays, as much that is grown in outside districts, and of an inferior quality, is shipped to Mocha and other leading districts, and re-shipped under the brands of the best products from those places.

Little is known of the early history of coffee, although we read of its being used as a beverage in Ethiopia as early as A. D. 875. At a more modern period, we note its introduction into Arabia from Africa—in the fifteenth century—and in Venice in 1615, and in England in about 1640. It was first introduced into Java by the Dutch between 1680 and 1690.

BORNEO.

This great island, whose area exceeds 284,000 square miles, lying on either side of the equator, between latitude 7 deg. 10 min. north and 3 deg. 40 min. south, and between longitudes 109 deg. 30 min. and 118 deg. 30 min. east, is the third in size among the islands of the Pacific.

The population is about three millions. There are many beautiful bays and inlets along its two thousand miles of coast line, although navigation is made exceedingly dangerous by the many islets and rocks that dot the sea along its shores. Beautiful rivers traverse Borneo, winding through its valley and plains, and are in most cases broad, navigable streams. Forty of this character are already known.

Great ranges of mountains rib the island here and there, some of them towering nearly 14,000 feet above the level of the sea.

TOPOGRAPHY.

Physically speaking, Borneo may be described as one immense forest, generally of moderate elevation

—that is, 300 to 700 feet—traversed by great rivers, which descend from a central group of mountains, and surrounded by wide alluvial plains, edged with mangrove swamps, or broken up into low deltas, constantly subject to inundation. It has, therefore, a physical character distinct from that of Java or Sumatra. Its plains are of much greater extent, and its mountains, on an average, do not attain the same elevation.

From northeast to southwest extends a chain of mountains, nearly parallel to, but at a great distance from, the west coast, which, in or near latitude 3 deg. north, curves around, to terminate at Cape Sipang. From this chain a short spur projects, and links it to a double range of lesser height, one of which runs southwest to a point near Cape Sambas, while the other pursues an irregular southeastern direction and reaches Cape Salatan. The culminating point of the first-named chain is Kinibulu, 13,680 feet in height. This is the loftiest summit on-the island, and on the east side of it lies a great lake, the source of numerous rivers.

The other important peaks are Kamangting, in the southwest chain—6,500 feet; Lunangi, in the southeast, 6,300 feet; Meratoo, also in the southeast, 4,000; Batang-Loòpar, east of Sarawak, 4,000; Krimbang and Saramboo, both south of Sarawak, 3,250 and 3,000. respectively; and Santibong, at the mouth of the river Sarawak, 2,050 feet. Thus it is evident that the general elevation of the island is not considerable. If it were sunk five hundred feet, at least four-fifths of its area would disappear, leaving several long peninsulas, of tolerable breadth, divided by broad ocean channels, and relieved by solitary mountain peaks rising here and there above the waters. If sunk one thousand

feet, nothing would remain but a few of these peninsulas ; the ocean ways would be broader, and the
mountain peaks wider apart.

We come now to the rivers of Borneo. In most
countries the configuration of the surface is determined
by the course of one principal river, or it is defined by
the basins of two or three main streams. Thus, Germany is marked out by the basin of the Rhone and
Loire ; Egypt, by the valley of the Nile. So far as our
knowledge of Borneo at present extends, it offers us
no such assistance in surveying and laying down its
superficial area. Its rivers are mostly tidal, but their
basins seem to be very narrow, and they descend lan-
guidly and slowly through vast level deltas, which
merge into inundated plains.

The littoral or shore country on the north and
northwest, a comparatively level tract about six hundred miles in length, is watered by a perfect network
of rivers, though probably not one of them exceeds a
hundred and fifty miles in its full career. They rise
from the range of mountains of which Kinibulu is the
culminating summit, and their course being short, are
more rapid than those in any other part of the island.
Some of them preserve their fresh water character
down to the very coast.

Tracing them from the north, we may notice, first,
the River Brunai (Borneo), a broad sheet of water,
navigable for some distance by large ships. Next, the
Binbula and the Judal, both of which are considerable
streams. Passing Cape Sinik, we observe the mouths
of the Rejang, which, at eighty miles from its mouth,
is one mile wide. Still larger than these is the noble

Butong-Lupai, which measures nearly five miles across, and can float a large frigate. The Sarawak, famous in the annals of English enterprise, is not so remarkable for its length or breadth as for its numerous branches, which ramify in such a manner as to afford to an extensive district all the advantages of water communication.

South of the equator we find the Mejak, the Sambas and the Kapooas. The first named was ascended by a Dutch steamer, as far as Malu, in March, 1855. The last named is one of the chief rivers on the island —perhaps the chief—measuring not less than seven hundred miles in its sinuous course.

On the south coast we notice the Djelli, the Pembuan, the Medawi, the Great Dayak, the Little Dayak, the Kahajau, the Murong, and the Bangermassin, or Burdo. This last is connected by several arms with the Murong on the west, and thence again with the Kahajau; so that a water-way penetrates into the very heart of the interior. In the lower part of its course it is continually overflowing the country, as its name indicates—*Bangermassin* ("frequent floods"). In the upper part it is called the *Dooson*, or village river, because its banks are occupied by several agricultural communities. It is fed on the east by the Nagara, a river which in itself is of considerable importance.

On the east coast the rivers are not so large nor so numerous, but we notice the Kooti, with its wide delta, extending over one hundred miles of coast. It was ascended by Major Muller, a Dutch officer, in 1825, and he had succeeded in crossing the mountains and descending into the valley of the Kapooas, when he was murdered by the Dyaks. Further to the north lies the *Pautai*, or river of Beron.

(Adams's Eastern Archipelago.)

VEGETABLE AND ANIMAL KINGDOM.

The soil of Borneo is very fertile, producing all that has made Sumatra and Java so famous. The flora is extensive and varied, the forests abounding in all the valuable woods and plants of the tropics, while the cultivation of the rattan, bamboo, banana, betel nut, cocoanut, bread-fruit, sugar-cane, tobacco, cotton, lemon, orange, clove, rice, nutmeg, ginger and opium poppy are but a portion of the valuable products. The sago and date-palm, the ebony, gutta percha, toa, tomano, prima vera, sandal, camphor and cinnamon trees adorn the forests.

The animal and mineral kingdoms are well represented; the former embracing the elephant and hippopotami, the rhinoceros, tiger and panther, the ourang-utan and the different species of the monkey tribe, roam through the vast forests or prowl among the jungles. In the latter kingdom we find gold, silver, lead, antimony, tin, iron and coal. The beds of many of the streams teem with that valuable gem, the diamond, mining for which has formed one of the industries on this island for ages.

Nor are the reptilian, finny or feathered species without an extensive representation. The swamps, morasses and forests are the homes of the great python, descending the scale through numerous species to the little coralilla, whose bite is certain death. The seas, rivers and bays teem with fish of all the species known in the tropics. Birds of the most beautiful and valuable plumage abound in the forest, while an endless variety of the aquatic kind frequent the pools, lakes and rivers.

DIAMOND MINING

One of the most valuable industries on the island of Borneo is diamond mining—a business followed in some countries for ages past. Borneo is not alone in her diamond-fields, as Sumatra, Australia and Tasmania have furnished some valuable gems. One found in the southwestern portion of Borneo, in the district of Mattan, and now in possession of the rajah of that region, weighs 367 carats, and is valued at something over $1,000,000.

Golconda, a district between Cape Cormorin and the Bay of Bengal, has been a celebrated diamond-field for ages past. Tavenier described a gem found in this region and taken possession of by the Great Mogul, as weighing 900 carats.

The diamond-fields of Brazil, located in the Sierra de. Frio, in the province of Minas Geraes, were discovered in 1728. A gem found here, and now belonging to the king of Portugal, weighs 1,680 carats, valued by some experts at the modest sum of $28,-000,000. As a carat in diamond weight is equal to the 150th part of an ounce Troy, and nearly the 137th part of an averdupois ounce, we have in this diamond a gem weighing nearly a pound Troy, and about four-fifths of a pound averdupois.

Brazil was for many years the principal diamond mining country, furnishing stones of great beauty and in great numbers to the world. In 1868 they were discovered in South Africa, where the district as far as known contains an area of 17,000 square miles. Many of the diamonds from this locality are of a yellowish cast, and not near so valuable as those found in other countries. The largest stone found here was the Stewart, weighing 288¾ carats, and of fine quality.

They are found in many other countries—in the Ural Mountains; in Hindostan; and in the United States, in North Carolina, Georgia, Virginia and California.

In addition to those already mentioned, the fame of the Kohinoor, of England, weighing 279 carats; the Orloff, of Russia, 195 carats; the Regent, or Pitt. 136¾ ; and the Sanci, 106 carats, is world wide.

Previous to the 15th century the gems were worn in the rough, just as they came from the mines, and of course lacked the brilliancy given to them by cutting and polishing. This art was discovered by Louis von Bergnen, in the above century, and gave to the diamond a value unequalled by any other gem.

Its uses in glass-cutting and in the manufacture of diamond-drills, for mining purposes, are so well known as to require no description here.

In mining for diamonds similar processes to those in use in placer mining for gold are resorted to. They are found just below the later alluvial deposits, intermixed in the stratum of gravel, clay and rolled quartz lying over the bed-rock, once forming the beds of streams and gravel deposits. From this deposit the stratum is washed through sluices with an abundance of water, the diamonds being found among the heavier particles remaining in the sluices after the washing.

The standard for valuing diamonds, presuming that they are of fair quality, is to multiply the square of the weight in carats by the value per carat. Taking the Kohinoor, for example: weight 279 carats, squared, would equal 77,844, which multiplied by the value, assuming it to be $20 per carat, would give $1,556,820.

This amount would be its presumable value in the weight given, although that was reduced by cutting and polishing to 186, and by still another cutting and polishing, which brought the weight down to 106 carats.

CHAPTER VII

ISLANDS.

A fleet descry'd
Hangs in the clouds, by equinoctial winds
Close sailing from Bengala, or the isles
Of Ternate and Tidore, where merchants bring
Their spicy drugs.

MILTON (*Paradise Lost.*)

SUMATRA.

EXTENDING in an oblique direction, to the north-west, lying almost immediately under the equator, running from latitude 6 deg. 10 min., south to 5 deg. 40 min. north, and between longitudes 95 deg. 10 min. and 107 deg. 10 min. east, is located the island of Sumatra. Twenty to thirty islands along the greater one's shores could be enumerated, but are of no special importance at present. Next to Borneo in size, having an area of about 160,000 square miles, with 4,500,000 people, Sumatra is a garden-spot, unsurpassed in valuable productions, except perhaps by Java.

Its position is easily remembered. Its northern portion is separated from the Malayan peninsula on the east by the Strait of Mallaca; on the west it is bounded by the Indian Ocean; on the south it is

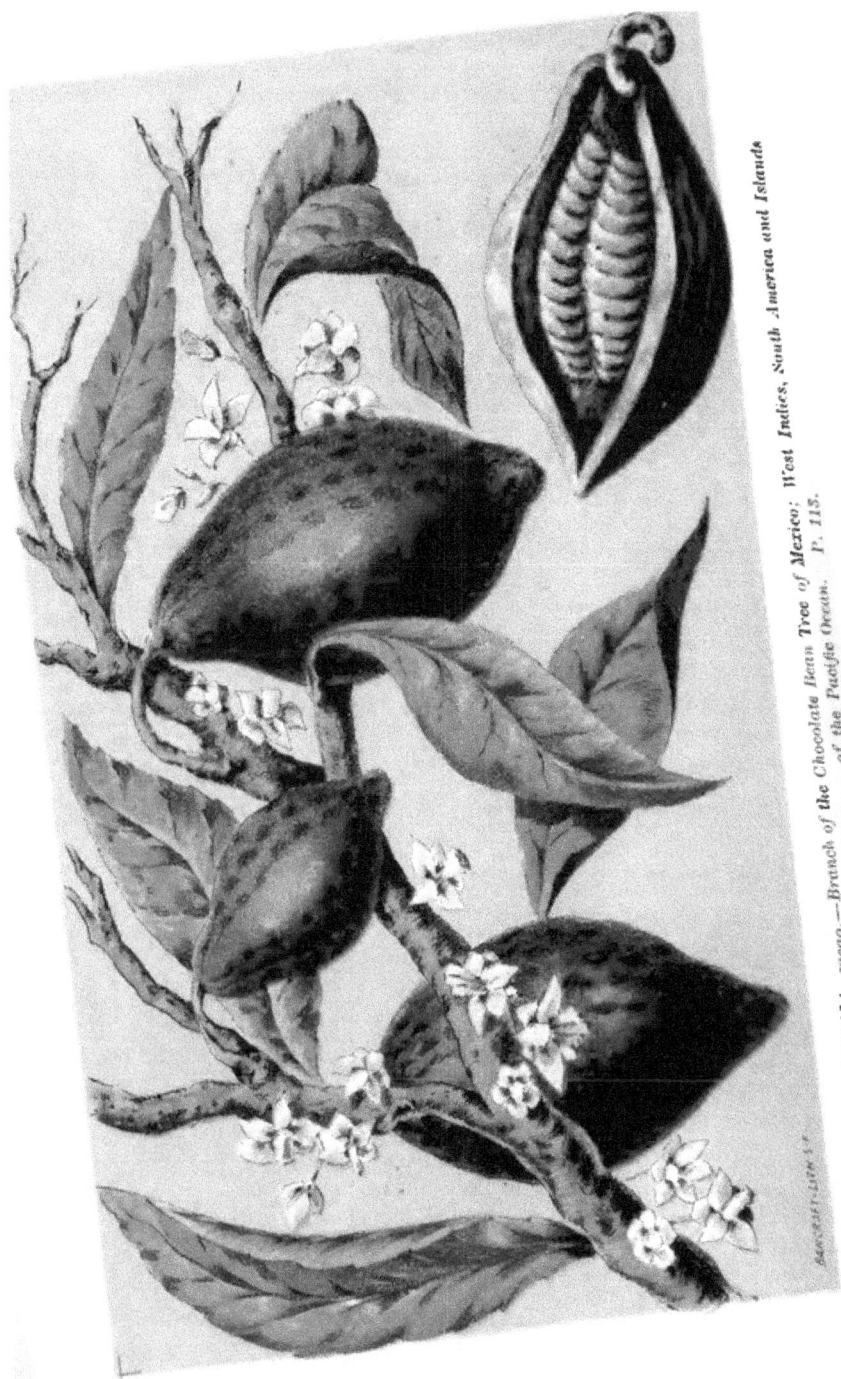

THEOBROMA, cacao.—Branch of the Chocolate Bean Tree of Mexico; West Indies, South America and Islands of the Pacific Ocean.   P. 115.

divided from Java by the narrow arm of the sea called the Strait of Sunda.

TOPOGRAPHY.

The eastern portion of the island is remarkable for its continuous levels, which are freely watered by several large but sluggish rivers—the Rawas, the Jambi, the Indgari—that form extensive deltas at their mouths, and have for ages been contributing to fill up the shallow sea, into which they fall. Very different in character the western portion. Here, from northwest to southwest, stretch range upon range of mountains, all running parallel to the coast, and increasing in elevation from 2,000 to 5,000 feet. These are broken up by short lateral valleys, and again by extensive longitudinal valleys, clothed with the fig and the myrtle, the arica and nibon palms. The littoral belt, or shore-land, varies greatly in breadth. On the southwest side of the island the mountains seem to start up directly from the ocean, and for nearly 400 miles the distance between the beach and the wooded base of the hills is two miles, though towards the north it widens on the average to six miles, and at a few points to twelve miles.

ANIMAL LIFE.

The reader will easily understand that the scenery in the western division of the island presents many romantic features. The mountain peaks rising so abruptly from the shore, and clothed with hanging woods, are necessarily objects of much grandeur; and intersecting valleys, enriched with a tropical vegetation, the forms and colors of which have a rare at-

traction for the eye of the traveler, are characterized
by numerous landscapes of great splendor. The interior of the island is but imperfectly known; but one
of these valleys, stretching up to the foot of Mount
Merapi, is fully 100 miles in length, and is regarded
by some authorities as the original home of the
Malayan race. Birds of bright tinted plumage dart
in and out of the thick boughs of the wide-spreading
woodland, and blend their voices, often harsh and
shrill, with the murmur of falling streams. Here in
the virgin forest the agile monkey leaps from branch
to branch; or the siawang, with his immense long
arms, five feet six inches across in an adult about
three feet high, swings himself with wonderful rapidity
from tree to tree. Here, in the remote recesses,
the ourang-utan live its melancholy life; the rhinoceros wades in the shallow streams, and the elephant
crashes through the jungle with colossal bulk. ⁂

FLORA.

Turning to the vegetable wealth of this great
island, we meet with the most valuable productions of
the tropical world. In the forest the huge trees, colossal in girth and of noble height, are linked together and surrounded by innumerable parasites and
creeping plants, often of great beauty, which inter-
lace with one another so as to form an almost impervious labyrinth. On the shore we meet with the
spreading mangrove, its pendulous roots closely
matted and intertwined, forming an incomparable
breakwater, and stemming the aggressive tide. Retaining the particles of earth that sink to the bottom
between them, they gradually, but surely elevate the

level of the soil, and as the new formation rises and broadens, a thousand seeds are sown upon it, a thousand fresh roots descend to strengthen and consolidate it; and in this way the mangrove repels the wave and asserts the supremacy of the land over the baffled sea. * * *

On the mountain slopes, from an altitude of five hundred to that of six thousand feet, the forest is largely composed of oaks of several species. They are noble trees, and of much value ; but in a commercial sense a higher value attaches to the *Dryanobalops,* which yields the all-important camphor. About one degree below the equator, its place is occupied by the *Diptuocarpus,* a tree of gigantic proportions, which produces the resin called "dammar."

On the rough bark of many of the forest trees grows that extraordinary parasite, the *Rafflesia,* the largest known flower, measuring fully three feet in diameter, and expanding a calyx which is capable of holding six quarts of water.

The principal exports of Sumatra are capsicums, ginger, betel, tobacco, indigo, cotton, camphor, benzoin, cassia or common cinnamon, rattans, ebony, sandalwood, teak and aloes, ivory, rice, wax, and edible birds' nests. To the list of the island products must be added rice, maize, sweet potatoes, taro, banana, mango, durian, pawpaw and citron. But even this enumeration gives but a faint idea of the variety and extent of its natural treasures.

CLIMATE.

Its climate is well adapted to the growth of so luxuriant a vegetation. Lying directly under the equator, the island enjoys great equability of temperature, the thermometer seldom falling below 76 deg. or rising above 93 deg. The constant rains brought up by the southeast monsoons counteract or mitigate the prevailing heat. In the highlands and mountain districts the climate is healthy, and the natives attain a considerable longevity; but in the low ground along the coast, and in the neighborhood of the mangrove swamps, Europeans, at least, drag on a sickly existence, and malaria exercises its deadly ravages.

The principal cities are Padang (the capital), Bencoolen and Palambang.

INHABITANTS.

The inhabitants of Sumatra are mostly of the great Malayan family, but in the north they seem to have intercrossed with the Hindus, and are distinguished by their strength, their stature and their fierce courage. The Chinese are numerous on the east coast. North of Menangkabu, where the pure Malays reside, live the Battahs or Batakhs, whose exact relation to the Malay it seems impossible to determine. They approximate, in many respects, to the Caucasian type, with fair complexion, brown or auburn hair, well-shaped lips and an ample forehead. All the natives of Sumatra, with the exception of some inland tribes, profess a modified Mohammedanism.

In Sumatra we find about fifteen volcanoes, four of which—Dempo (10,440 feet), Indrapura (12,140), Talang (8,480), and Merapi (9,700 feet)—are of considerable importance; the others do not exceed six or seven thousand feet in elevation.

(Notes from Adams's Eastern Archipelago.)

SINGAPORE.

This little island, located between latitude 1 deg. and 1 deg. 32 min. north, and longitude 103 deg. 30 min. and 104 deg. 10 min. east, has long been celebrated for its many valuable products, being more widely known than almost any other island in the East. Situated at the eastern extremity of the Straits of Mallacca, it has long formed the distributing point for the products of these regions.

The town of Singapore has about 100,000 inhabitants—Malays, Hindoos and Chinese—and is located a mile or so back from the straits, in the mouth of a river; the freight to and from the town being handled by lighters.

The island itself has an area of about 220 square miles, and is surrounded by about fifty small islets, of no great commercial importance in the past or present as distributing points, yet the fisheries, the turtle, tortoise and *beche de mer*, found on some of these little desert spots, are considerable. The whole area, including the islets, may be estimated at 400 square miles. The British hoisted their flag over Singapore in 1819, but it was not till 1824, when the main island, with the adjoining isles located within ten miles of the shores of Singapore, were ceded to the East India Company by the Malayan princes, that Singapore sprang into commercial importance.

The Straits of Mallacca narrow down at one point to a quarter of a mile in width between the island and the Malayan Peninsula. In some respects this is unfortunate for the inhabitants of Singapore, as one of the favorite methods of the tiger, the great man-eater of the East Indies, is to swim this channel from the mainland and make a meal off of a native. It has been estimated that Singapore loses one inhabitant a day in supplying this demand.

.

CELEBES.

Between the parallels of latitude 1 deg. 45 min. north and 5 deg. 52 min. south, and the meridians of longitude 118 deg. 45 min. and 125 deg. 17 min. east, lies an island of the most extraordinary configuration, which some writers compare to a tarantula spider, others to a couple of horse-shoes joined at the fore parts. Neither comparison is very accurate. It consists of four long peninsulas—the largest being the northernmost—of which two are directed eastward, with a deep gulf between them (the Tomini Gulf), and two others southward, with the Boni Gulf separating them from each other, while the first of the two is separated from the second of the other two by the Tolo Gulf. These four peninsulas project from a narrow neck of land which runs due north and south.

The peninsula of Menado, the first of the four peninsulas, sweeps north, then east, and lastly northeast, with a length of 400 miles and a breadth of 12 to 60 miles. That of Bulante, east, is 160 miles long and from 30 to 95 miles broad ; the southeast peninsula is about 150 miles by 30 to 90 miles ; and the southwest (that of Macassar) forms a tolerably regular parallelogram, 200 miles long and 65 miles broad. They are all formed of mountain masses, and describe a kind of backbone, 150 miles long and 105 miles broad.

The Gulf Tomini or Gorontala, on the northeast, is 240 miles long, and from 55 miles at its mouth it broadens, as it strikes inland, to fully 100 miles ; that

of Tomaiki, or Tolo, on the east, is of ample dimensions
at its mouth, but narrows towards its upper extremity;
and that of Macassar or Boni, on the south, is probably upwards of 200 miles in length, with a width varying from 35 to 80 miles.

Apart from these conspicuous indentations, the
coast line is broken up by numerous bays, such as
those of Meuado, Amoorang, Kwandang and Tontoli,
on the north; Palos and Panepane on the west, and
Bulante, Tolowa, Nipa-Nipa and Staring on the east.

To sum up, we have an island of Celebes, 150
miles long and 105 miles broad, throwing off four peninsulas of varying magnitude; the superficial area of
the whole island being estimated at 71,791 square
miles.

We might conjecture that an island so exposed
to the sea breezes would be visited by abundant moisture, and being included in the tropic zone, and immediately under the equator, would necessarily present a
vegetation of remarkable richness and variety. Such,
indeed, is the case, and Celebes has fair claims to be
regarded as the loveliest and most bounteous of all
the islands of the Eastern Archipelago. Its scenery
combines every charm that can gratify an artist or in-
spire a poet; it has the immense forests of Corneo and
the meadows and vales of England; the exuberant
wealth of the tropics, and the gentleness and grace
that distinguish the regions of the temperate zone.

Broad rivers, lofty heights, far-spreading woods,
deep, bowery hollows, immense breadths of fragrant
greensward—it has all these, mingled with rare and
beautiful forms of vegetation, and enlivened by glorious displays of color, which give to each bright, strange
landscape an individuality of its own. To all this add

a fresh and healthy climate, which neither enfeebles the
mind nor undermines the physical health, and it may
be conceded that Celebes is an enchanted land.

(Adams's Eastern Archipelago.)

SANGIR GROUP.

North of Celebes, between latitude 2 deg. and 4
deg., is the Sangir group, about fifty in number, with
an area of 1,500 square miles, and a population of
30,000. Like many of the islands and groups in these
seas, they are afflicted with the eruptic volcano, whose
destructive ravages are to be seen on every hand. At
Great Sangir, the largest island of the group, having
an area of some 300 square miles, we find, in the
northwest portion, the active volcano of Abu. In
March, 1856, a fearful outburst took place here; the
burning lava, boiling water, scoria and ashes laid
waste the surrounding country, destroying towns and
villages, sweeping over the fine plantations, leaving
all within reach a vast, burning, smoking waste.

If this were all to relate of this eruption, it could
be passed over with barely a glance; but when the sad
fate of three thousand people, who lost their lives,
caught in the burning lava or in floods of boiling water,
or smothered in clouds of sulphurous smoke and ashes,
is added, it darkens the history of these island regions
like a funeral pall.

This island group produces nearly all of the tropical products in the greatest abundance. With a fertile soil, made beautiful by an industrious people, they
appear like gems dotting the southern seas. But, like
the neighboring isles, they lie over the track of the
great eruptic fire-belt, whose terrible outbursts too fre-

quently devastate the lands and convulse the foundations of the deep.

MOLLUCCA ISLANDS.

The name Molluccas is employed in a restricted,
and also in a comprehensive or general sense. It is
applied, in the first place, to the Royal Islands, lying
off the western coast of Gilolo, and washed by the
Molluccas Passage, which separates Gilolo from Celebes. In a wider sense, the name Molluccas is applied
to all the islands or groups of islands lying between
Celebes and New Guinea. They are commonly divided, according to the three residencies, into the Ternate, Amboyna and Banda groups, which contain, respectively, the following principal islands :

1. *The Ternate Islands*, including the Molluccas
proper—comprehending Ternate, Gilolo, Batchian,
Obi, Mortui, and the Kaiva Islands ;

2. *The Amboyna Islands*, including Amboyna,
Ceram, Bouru, Goram, Amblau, and some smaller
isles ; and

3. *The Banda Islands*, including Great Banda or
Luthoir, Banda Neira, Pulo Run, Pulo Ai, Goenong
Api, Rosengyn, Kapal, Pisang, Spethau and Vronwen.

These numerous islands are all mountainous and
mostly volcanic, and their forms of animal and vegetable life exhibit but few and unimportant differences.
They may, therefore, be properly comprehended under the one general title of the Molluccas.

We shall visit them in the following order : Banda
and adjacent islands; Amboyna, Ceram, Bouru, Goram; and Ternate, Gilolo, Batchian and adjacent
islands. The inhabitants are Molluccan-Malays, and

their religion is principally Mohammedan. * * *
So much for the position of these charming islands,
which escaping the dry winds that blow over the
Australian deserts, are remarkable for their fresh
greenery and the plentifulness of their vegetation.

HISTORY.

They were first made known to Europeans by the
Portuguese navigator, D'Abreu, but the Chinese
and Arabs, and probably the Hindoos, had long previously included them in the range of their commercial enterprise. D'Abreu, according to the chron-
icler, DeBarros, had the assistance of Javanese and
Malay pilots who had made the voyage; and DeBarros
adds, that every year Javanese and Malays repaired
to Lulotain (that is, Great Banda) to load cloves, nutmegs and mace, for it lay in the latitudes most easily
navigated, and where ships were most secure, and as
the cloves of the Molluccas are brought thither by
vessels belonging to those islands, it was unnecessary
to go to the latter for the much prized spices. In the
five islands, says DeBarros, namely, Louthoir, Re-
sengyn, Pulo Ai, Pulo Run and Banda Neira, grow
all the nutmegs consumed in every part of the world.
He gives the then population as 15,000—a very much
larger number than at present, and further says of
them: The people of these islands are robust, with
lank hair and a tawny complexion, and are of the
worst repute in these regions. They follow the
sect of Mohammed, and are much addicted to trade,
their women performing the labors of the field.
They have neither king nor lord, and all their
government depends on the advice of their elders,

and as these are often at variance, they quarrel among themselves.

NUTMEG.

The land has no other export than the nutmeg. This tree is in such abundance that the land is full of it, without being planted by any one, for the earth yields without culture. The forests which produce it belong to no one by inheritance, but to the people in common.

For about a century the Portuguese monopolized the commerce of these islands, and throughout this period maintained a friendly intercourse with the natives. In 1609 the Dutch, however, resolved to annex them to their Eastern possessions, and invaded Great Banda with a force of 700 soldiers, but falling into ambuscade, were compelled to retreat with considerable loss. They then began a war of extermination, which was prolonged for eighteen years, and brought to a successful issue only through the efforts of a large expedition from Java, commanded by the Governor-General in person. In this prolonged struggle, the natives, who fought with great courage and resolution, lost 3,000 killed and 1,000 prisoners. The survivors fled to the neighboring islands, where they were merged in the general mass, so that scarcely a vestige of their language or customs is now known to exist.

LOUTHOIR.

Of the little island group of Louthoir, it is said that beneath the shade of the lofty kanary trees, deriving their nourishment from the thin but warm volcanic soil, and fed by the constant moisture, the handsome glossy-leaved nutmeg trees, twenty to thirty feet high, line the roads and bloom in the gardens and spread over all the open places. They are very fair to look upon, with their thick-spreading branches, the tallest sprays of which are fifty feet high. The flowers are small and yellowish. The fruit, before it is fully ripe, resembles a peach that has not yet been tinted with red; but this is only the epicarp, or outer rind, which is of a tough fleshy consistence, and on maturing splits open into two equal parts, revealing a spherical, polished, dark-brown nut, enveloped in crimson *mace*. In this stage it may be fairly described as the most beautiful fruit in the cornucopia of Pomona.

It is now picked by means of a small basket fastened to the end a long bamboo. The epicarp being removed, the mace is carefully taken off and dried in the sun, which changes its bright crimson to an obscure yellow. It is then ready to be packed in cakes and shipped to market. Next the nuts are spread on a shallow tray of open basket-work, and exposed for a period of three months to the action of a slow fire. By the end of that time the actual genuine nutmeg has so shrunken that it rattles in its dark-brown shell. The shell is broken, and the nutmegs after being sorted, are packed in large casks of teak-wood, which are duly branded with the year in which the fruit was gathered and the name of the plantation where it was grown.

AMBOYNA.

Mountains, hills, rocks, forests, noisy burns and rippling brooks, with well wooded valleys running in among the highlands and low fertile country stretch-

ing along the shore. Such is the general character
of Amboyna. It is not one of the fairest or richest
islands of the Archipelago; much of its surface is
bare and barren, and it presents but little of that exuberant vegetation which we are accustomed to asso-
ciate with the tropics. In fact, it owes its celebrity
and its wealth to one special vegetable product—the
clove-tree—(*caryophyllus Aromaticus*). Such being
the case, and groves of clove-trees, with their bright
green verdure, being the pleasantest objects in the
island, before we go further it will be well for us to
devote some attention to so remarkable a source of
wealth.

CLOVE.

We first hear of cloves in Europe about A. D.
175–180, in the reign of the Emperor Aurelian, when
they are mentioned as imported into Alexandria from
India—the Isthmus of Suez and the Red Sea forming
then as now the great highway along which flowed
the traffic of the East. They were carried by the
Javanese and Malays from the Molluccas to· the
peninsula of Mallacca; thence the Telingas, or
Klings, transported them to Calicut, the once famous
capital of Malibar. From Calicut they passed to the
western shores of India, and crossing the Arabian Sea,
found their way up the Red Sea to the Egyptian port.

The native name for this fruit is *chenki*, which
may be a corruption of the Chinese *theng-ki*, or
"sweet smelling nails." The resemblance to a nail
has also suggested the Dutch name, *krind-nagel*, or
"hub-nail" (the trees are *nagelen-boomen*, or "nail-
trees"), and the Spanish *clavos* (Latin *clavus*, a nail),
whence comes our English "clove."

The clove tree belongs to the order of Myrtles, which includes the guava, pomegranate and the rose-apple. Its topmost branches are usually forty or fifty feet from the ground, and the full-grown trunk measures eight to ten inches in diameter. It was originally confined, says Bickmore, to the five islands off the west coast of Gilolo, which then comprised the whole group known as the Molluccas—a name that has since been extended to Bouru, Amboyna and the other islands off the south coast of Ceram, where the clove has been introduced and cultivated within a comparatively late period. On these five islands it begins to bear in its seventh or eighth year, and sometimes continues to yield until it has reached an age of nearlg one hundred and fifty years ; the trees. therefore, are of very different sizes. Here at Amboyna it is not expected to bear fruit before its twelfth or fifteenth year, and to cease yielding when it is seventy-five years old.

A quaint description of this celebrated tree is given by Pigafetta, who accompanied Magellan in his voyage around the world: It attains a pretty considerable height, and its trunk is about as large as a man's body, varying more or less according to its age. Its branches extend very wide about the middle of the trunk, but at the summit terminate in a pyramid. Its leaf resembles that of the laurel, and the bark is of an olive color. The cloves grow at the end of small branches, in clusters of from ten to twenty, and the tree, according to the season, sends forth more on one side than the other. The cloves at first are white, as they ripen they become more reddish, and blacken as they dry.   *   *   *

To this we may add that the buds when young

NATIVE LUXURY IN THE MOLUCCAS.

are white, afterwards they change to a light green, and finally to a bright red, when they must at once be gathered, which is done by picking them by hand, or beating them off with bamboos, so that they drop in showers on cloths spread beneath the trees. When they have been dried in the sun—a process which changes them from red to black—they are ready for market. The gathering seasons are from June to December. The soil best adapted to the tree seems a warm, loose, sandy loam.

CHOCOLATE BEAN.

Another of the valuable products of this group, as others of the Eastern Archipelago, is the *cacao theobroma*, the chocolate bean of commerce. It is not native here, but is one of the few things which the Orient has borrowed from the West. The Spaniards discovered it in Mexico, and transplanted it to their settlements in South America and the West Indies. Thence it traveled to the Molluccas. It is also cultivated in Guinea and Brazil.

The cacao tree seldom exceeds twenty feet in height. Its leaves are large, oblong and pointed ; its flowers hang in pale red clusters, not only from its branches, but also from its trunk and roots. Hence a cacao plantation has a singular and striking appearance, as Humboldt did not fail to notice. Never, he says, shall I forget the profound impression made on my mind by the luxuriance of tropical vegetation when I first saw a cacao garden. After a damp night, large blossoms of the *theobroma* ("drink for gods!") issue from the root a considerable distance from the trunk, emerging from the deep black mold. A more

*8

striking example of the expansive powers of life could hardly be met with in organic nature.

The fruits are large, oval-pointed pods, about five or six inches long, and divided into five lobes or compartments, containing from twenty to forty seeds, the cacao of commerce, enveloped in a white pithy substance.

In localities well sheltered from the wind the grower sows his seeds. In two years the plant attains a height of three feet, and throws off numerous branches, all of which are removed, with the exception of four or five. In the third year the fruits appear, but the tree does not yield fully until six or seven years old, after which it produces abundant crops for upwards of two decades.

When the pods are first picked they are remarkable for a peculiar pungency, which can be converted into the highly valued aromatic principle only by a process of fermentation. Therefore they are thrown into pits, covered with a thin layer of sand, stirred at intervals, and allowed to remain for three or four days. After which they are taken out, cleaned, dried in the sun, packed in cases or sacks, and dispatched to the market. They are best known in Europe in the form of chocolate, being roasted, ground into a smooth paste and flavored with vanilla or other spices.

The pineapple, too, is found in this, as well as on the islands of adjacent groups.

SAGO PALM.

At Ceram Island, the largest of the Molluccas, one of the chief natural productions is the sago palm, known in botany as the *Sagus Lœvis* and *Sagus Rum-*

phii. It is not only more plentiful here than in any of the adjoining islands, but attains to greater perfection. It grows to the height of one hundred feet, and a single tree will sometimes yield twelve hundred pounds of starch, instead of four hundred pounds, as at Amboyna. The tree, in its early stage, is very slow of growth, but when it has once formed its stem it shoots up rapidly, and assumes its crown of far-spreading foliage and colossal efflorescence. Before the flower ripens into fruit the tree must be felled, as otherwise the farina which man uses for his food would be exhausted.

The sago, which forms so important an article of commerce, is prepared from the soft inner portion of the trunk, the latter being cut into pieces about two feet long, which are then split in half, and the soft substance scooped out and pounded in water till the starchy matter separates, when it is drained off with the water, allowed to settle, and afterwards purified by washing. The substance thus obtained is sago meal ; but before being exported to the European markets, it is made into pearl sago by a Chinese process carried on at Singapore. The rough meal is subjected to repeated washings and strainings, then spread out to dry, and broken into small pieces, which, when sufficiently hard, are pounded and sifted until they are tolerably uniform in size. Small quantities, finally, are placed in a large bag, which is suspended from the ceiling, and shaken backwards and forwards for about ten minutes, until the sago becomes pearled or granulated, after which it is thoroughly dried and packed for exportation.

(Adams's Eastern Archipelago; Bickmore's Travels in; Wallace: Malay Archipelago.)

CHAPTER VIII.

ISLANDS

The winds are aw'd, nor dare to breathe aloud,
  The air seems never to have borne a cloud,
Save where volcanoes serd to heaven their curl'd
  And solemn smokes, like altars of the world.
                                   EDWARD C. PINCKNEY.

NEW GUINEA.

NEXT to Australia in size, probably—lying just to
the north, and separated from it at one point by
the narrow Straits of Torres—is New Guinea. It
was discovered in 1511 by Antonia d'Albreu and
Francisco Serram. The population is altogether native, and numbers fully 500,000. The area is about
300,000 square miles.

The interior is wholly unknown to Europeans,
and our acquaintance even with the coast line cannot
be described as complete. The island is, however,
most irregular in form. On the west a deep basin,
called Geelvink Bay, sweeping inland from the north,
almost meets the Gulf of McClure, entering from the
west, and so forms a bold and extensive peninsula
connected with the mainland by a very narrow isthmus.

There is reason to believe that the island is very

mountainous, with deep, well-wooded valleys breaking
up the various chains, and with meadow lands extending from the base of the mountains to the sea. The
summits of the southern peninsula attain a far loftier
elevation than those of Australia. Mount Owen Stan-
ley, for instance, is 13,205 feet high, and Mount Obru
is 10,200 feet. A magnificent chain follows the line
of the north coast with much faithfulness, forming the
ranges of the Cyclops, which terminate in the Island
of Jobi; and further west, of the Arfak and Amberba-
kin, with a maximum height of about 9,000 to 9,500
feet. On the southwest the limestone formation crops
up in terraced heights, which rise one above another
like the stages of an amphitheatre, until they mount
above the snow line; the warm and humid forests of
the tropics lying at their base, their crests uprearing
the icy, snowy pinnacles of an Arctic world. The
Snow Mountains are 15,400 feet above the sea-level.

Valley and plain and hill, ravine and mountain
steep, all are clothed with a vegetation that almost de-
fies description by its luxuriance and variety. When
the island has been thoroughly explored, we may expect to hear that it is not inferior to Java or Borneo in
fertility of soil. It is certain that it produces all the
richest of fruits and the most valuable growths of tropical nature. In the lowlands, bread-fruit, cocoanut,
banana, sago, betel, orange and lemon, and a multitude
of other luxuries; in the higher grounds, magnificent
forest trees, the kanary, the masool, the wild nutmeg,
ebony and iron wood. Sugar cane, tobacco and rice
yield abundant crops; maize and yams are also cultivated, and among the glories of the forest is the camphor tree.

Nor is the usual parasitical exuberance wanting;

epiphytous plants overarch the wooded glades, and
creepers of every description hang in festoons from
bough to bough. Among the wealth of leaf and bloom
the paradise birds build their sequestered nests, and
the echoes ring with the shrill cries of parrots and lo-
ries, and the murmurs of carpophagous pigeons.

Animal life is not so abundant as vegetable. The
mammals are few in number, and most of them are
marsupials of the Australian type; though New Guinea
possesses some indigenous species of kangaroos, and
more particularly two species which are strictly arborial
in their habits. Wild swine are plentiful, as also the
wood-cat. Of birds, about sixty species have been
particularized. Insects astonish by their numbers, and
dazzle by their brilliancy of coloring. The rivers swarm
with fish, and so do the surrounding seas.

The great island is not alone in her grandeur, for
along her shores, and no great distance from the main
land, there are at least one hundred islands. The area
of these would probably reach 10,000 square miles,
with a population of over 20,000 people. The physical features, as well as products, are similar to those
of New Guinea.

Curiously, the main island, with those lying close
to it, if we except a very few of the Molluccas, are the
homes of that most beautiful of birds, the Bird of Paradise (*Paradisaidæ*). They are not to be found on
any of the other islands. Of the Paradisæ, twenty
species are already known; their beautiful plumage
being much sought after to supply the fashionable
markets of the world.

ADMIRALTY ISLANDS.

About two hundred miles to the northeast of

Papua are the Admiralty group, about thirty in number, with something over 1,000 square miles of area, and a population of 25,000.

They are not a prominent group in a topographical sense, lying but a hundred feet or so above the sea level; although for fertility and indigenous tropical products, they rival some of the more famed islands. The inhabitants are very similar to those of New Guinea.

Basko, or Admiralty Island, is the principal in the group, having an area of about 450 square miles.

They were first discovered by the Dutch navigator Cornelius Schooten, in 1616, and were afterwards re-discovered by Phillip Carteret, in 1767, who located them definitely on the charts, and gave them the name they bear to-day.

NEW IRELAND.

South by east from the Admiralty group, and northeast from Papua, we have New Ireland and New Britain.

There are some six islands in the former group, New Ireland being the only one requiring any description here. It is about two hundred miles long by fifteen wide, with some hilly ranges rising to a height of 2,000 feet. The island is well wooded and watered, and said to be healthy in the extreme. Tropical fruits are to be found in great abundance; while the forests that cover the sloping hills from valley to summit, abound in fancy woods of great commercial value. There are great numbers of tortoise taken here, whose shell is of the most beautiful and valued kind. The inhabitants, like all the islands around New Guinea, excepting always Great Australia, are of the woolly-headed negro type, and may be set down at 16,000 in number, the whole group giving an area of 4,300 square miles.

NEW BRITAIN.

Southwest from New Ireland, not many miles away, lies the New Britain group, inhabited by 20,000 people, and having an area of 10,500 square miles, within the limits of the eight islands. They have the same physical features as New Ireland, with a richness of tropical vegetation unsurpassed in these latitudes. The natives are of the Papuan type, uncivilized and treacherous in the extreme. The products, are like those of the islands in the surrounding seas. This group, like many others, needs but to be touched by the magic wand of enterprise and civilization to place them among the important islands of the world.

LOUISADE ARCHIPELAGO.

Southeast from Papua, stretched over 350 miles of the sea, the Louisade Archipelago lies, a long, low group, with scarcely any prominence in the way of hills and mountains. Little is known of the value of the products of these islands, the fierce and treacherous disposition of the natives preventing close commercial relations. As far as known, the area of the group does not exceed 1,500 square miles; while it is safe to estimate the population at 5,000. The islands have every appearance of being very fertile, tropical verdure spreading over the cluster on every hand. The natives are negritos of the worst type.

PHILLIPPINE ISLANDS.

This remarkable group of islands, numbering over 1,200, with an area of about 150,000 square miles, and a population of 5,000,000, is said to have been discovered by Magellan in 1521. This, like many of the modern discoveries, and credited to particular discoverers, will not bear the test of research; as Marco Polo sailed through the group as early as the thirteenth century.

For a description of this great island cluster, with some of their products, I am indebted to "Adams' Eastern Archigelago," and the writings of that great navigator and bold buccaneer, William Dampier.

These islands present so many interesting and attractive features, that we shall attempt a somewhat detailed account, beginning with a general view of their prominent characteristics, glancing at the history of their discovery by Europeans, and concluding with some sketches of their scenery, and leading forms of animal and vegetable life.

The principal islands are Luzon, Mindanao, Mindoro, Samar, Panay, Leyte, Zebu, Negros, Bohol and Alawan. The whole cluster is divided into groups; the Sooloos, Bissayas, Pasay, Bashu and Babwyan being the most important.

TOPOGRAPHICAL.

The larger islands of the group appear to produce a powerful impression on the imagination of the voyager, to judge from the many glowing pictures contained in various narratives. Their coast line is bold and irregular, broken up by numerous romantic headlands, the declivities of which are green with abundant foliage; by long, narrow tongues of land, with forest growth extending to the very margin of the sea; by broad bays, each capable of accommodating an imperial fleet; and narrow inlets and creeks, so embowered in shade that large ships might harbor in them and not be discovered by a passing enemy. Then, from the bright and picturesque shore, the ground rises inland with a continual ascent, until the undulating plains are succeeded by low ranges of wooded hills, and these by lofty ranges, which here and there culminate in magnificent mountain peaks. In and among these ranges, which are irregular in their direction, and throw off numerous short chains and spurs, lie slopes of perennial verdure, and valleys so gifted with the bounties of nature that they surpass the dreams of the Arcadian poets. Here, too, are broad, deep lakes, in their general features reminding the traveler of the charming basins of the Scottish Highlands; while many streams flow through the verdurous glens to unite in ample rivers, which, with full channels, descend to the sea.

The vegetation of the Phillippines is among the richest of tropical climes. A fertile soil is assisted by a genial climate. Droughts are unknown; the tropical heats are tempered by abundant moisture and by the constant alternation of the land and sea breezes. In the western portions of the group, the rainy season begins in June and ends in September; in the east it begins in October and ends in January; and the rains are then so heavy and so continuous that the low grounds are converted into extensive lakes.

This inundation, however, increases the fertility of the soil and favors the growth of exuberant crops. It may almost be said that the only misfortune to which the islands are liable—the only shade on a picture which astonishes us by its splendor—is the frequency and severity of their earthquakes. They form a part

of the great volcanic chain to which, in describing the
Eastern Archipelago, we have so often found it necessary to allude; and they possess several volcanoes,
both active and extinct—among the most important of
which is that of Taal. Manilla, the capital of Luzon,
and the chief town of the group, was ruined by a convulsion which broke out on the evening of the 3d of
June, 1863. The cathedral, with its noble dome, was
shattered into ruins by a shock which occurred while
the priests were chanting vespers. The Viceroy's pal-
ace was destroyed, and the British consulate. Not
one of the churches escaped, and the only one left
standing (that of Binondo) was rent from roof to base-
ment. Nearly two thousand persons perished.

MINERALS.

The Phillippines are not only rich in vegetation,
but abound in subterranean treasures. The sands of
their rivers yield no inconsiderable quantities of gold-
dust. All the palaces of earth might be rebuilt from
their extensive quarries of marble and limestone. The
coal fields cover a wide area and produce an excellent
fuel. Iron—the wealth of strong and powerful nations
—and copper of the best quality, are found in all the
mountain ranges. Sulphur, magnesia, quicksilver,
vermillion, saltpetre and alum are also plentiful. So
vast, indeed, are the resources of the Phillippines, that
only an able government is needed to give them the
position of a wealthy, influential and prosperous commercial state. But the colonial administration of Spain
has never been marked by either vigor or sagacity;
and though the recent development of commerce has
been considerable, it is by no means proportionate to
the capabilities of these beautiful islands.

The forest trees which cover the valley slopes and
ascend the mountain sides are very valuable. Among
the plants cultivated for use, we find the gornuti or
cabonegro palm, the abuca, the cocoa and other palms,
the pineapple, the cacao tree, cotton and coffee, the
tamarind, indigo and sugar-cane. Tobacco is largely
grown, and the Manilla cigars are scarcely less celebrated than those of Havana. Rice is raised in immense quantities, and forms a principal article of trade ;
and the vegetable wealth of the group also includes
cassia, cloves, red and black pepper, vanilla, cinnamon,
nutmegs, maize, wheat, yams, the sweet potato, and a
variety of the most delicious fruits on which the ripening sunshine of the tropics falls.

ANIMALS.

Animal life is neither less various nor less exuberant. The horses of the Phillippines are small, but
strong and lively ; the deer supply a capital venison ;
hogs, goats, sheep, buffaloes and oxen are bred by the
agriculturist ; foxes and gazelles frequent the valleys ;
monkeys, squirrels, wildcats, and the *bagua*, a kind of
flying cat, the woods. The jungles are enlivened by
the bright plumage of humming birds, parrots, and the
rhinoceros bird. The sea swallow builds her edible
nest in the hollows and caves of the rocky coasts. The
forests swarm with eagles, falcons, herons, pigeons,
game cocks, quails, and the lakes with aquatic birds.
Pools and rivers teem with fish ; but here an unplea-
sant fact obtrudes itself upon us—crocodiles are numerous. Serpents lurk in the dense growth of the
forests ; leeches swarm in the swampy lowlands ; rep-
tiles abound, and insect life displays itself with a luxu-

riance which both native and stranger find good cause
to lament.   *   *   *

We have spoken of the forest trees. They attract attention by their enormous bulk and by their
huge canopies of spreading foliage. They are bound
together by the remarkable bush-rope or palaseru,
which grows in festoons several hundred feet in length;
while a whole world of epiphytous plants, parasites,
creepers, climbers and liaries find nourishment in their
bark, or support on their stalwart arms, and spread
everywhere such a tangle of leaf and stem and blossom,
that the traveler can only force his way into the forest
depths, axe in hand.

INHABITANTS.

The industrial occupations of the natives include a
very ingenious method of working in horn; the manufacture of gold and silver chains; of cigar cases, and
fine hats in various vegetable fibres; of beautifully
colored mats, embroidered with gold and silver; the
dressing and varnishing of leather; ship-building and
coach-building. The manufacture of cigars gives employment to a large number of people. The cordage
of the Phillippines is held in good repute. The textile
fabrics are said to be fifty-two in number, from the
delicate and costly shawls and handkerchiefs, made
from the fibre of pine-apple leaves, called *piñas*, and
sold at the rate of one or two ounces of gold a piece,
down to a coarse cotton and stout sacking, wrought
from the fibres of the abaca and gornuti palms.

We have nearly completed our general view of
islands, but a few details seem wanting for the full information of the reader. The two principal races are
the Tagals and Bisayers, who inhabit the towns, vil-

lages, and cultivated lowlands, and are mostly Roman
Catholics, though a considerable number remain faith-
ful to the creed of Mohammed. In the mountainous
interior we find what is probably the original race, the
Oceanic Negroes, a black-complexioned, negroish
people, closely resembling in their persons and customs the Papuan Alfoories. They are chiefly heathens,
practicing a wild and crude idolatry, or otherwise observing no religious form at all, though not free from
degrading superstitions. Among the industrial population a foremost place must be given to the Chinese
immigrants, who, however, do not settle permanently
in the islands; while the Mestizos, or half-breeds, who
are mostly of Chinese fathers and native mothers, exhibit a remarkable degree of activity, enterprise and
industry. Spaniards are few in this Spanish colony,
except in the military and naval service.

DAMPIER.

Dampier visited these islands in 1686, as pilot on
board the *Cygnet*, a privateer, or buccaneering vessel,
commanded by Captain Swan. They were kindly received by the natives, though their piratical character
seems to have been suspected. They obtained a supply of fresh provisions; and Dampier for the first time
saw the bread-fruit tree, the staff of life to so many of
the Polynesian tribes. At the flying proas, or sailing
canoes of the natives, the visitors were greatly astonished. They were admirably built, and so swift that
Dampier was persuaded that one of them would sail
twenty-four miles an hour; and another had accomplished the distance between Guahan and Manilla, or
400 leagues, in four days.

Dampier describes the trees of Mindanao with some degree of particularity. In his time they were curiosities, and scarcely known to Europeans, even by repute; but now we are all familiar with the properties of the bread-fruit and the cocoa-nut, the nutmeg and the banana, the durian and the plantain.

HIS ACCOUNT OF THE PLANTAIN.

The plantain he boldly terms the king of all fruit. He will brook no rivals near its throne, not even the cocoanut palm, gracefulest of all vegetable wonders, which wins the admiration of every cultivated eye with its slender, shapely column and lifted crown of plumes. The tree that bears the plantain is, he says, about three feet or three and a half feet around and ten or twelve feet high. It is not raised from seed, but from the roots of old trees of the same kind. If these young suckers are taken out of the ground and planted in another place they will not fructify for fifteen months, but if allowed to remain in their own soil they will fructify in twelve. As soon as the fruit is ripe the tree decays, but several young ones are ready to take its place. On first emerging from the ground it springs up with two leaves, and by the time it is a foot in height two more spring up inside the first couple, and shortly afterwards two more within *them;* and so the brave work goes on. By the time it is a month old a small stem about the size of a man's arm is discernible, as well as eight or ten leaves, some of which are four or five feet high. The first leaves, however, are not more than twelve inches long and six broad; the stem that bears them is no bigger than a man's finger, but the leaves increase in size as the tree increases in height. The old leaves spread off as the young spring on the inside, and their tops droop downwards, being of a greater length and breadth in proportion as they are nearer to the roots. At last they decay and drop off, but the young leaves always blooming at the top preserve the green and flourishing aspect of the tree. * * *

Thus the body of the tree seems to be made up of many thick skins, growing one over another, and when it is full grown, out of the top springs a strong stem, harder in substance than any other part of the trunk.

This stem shoots forth at the heart of the tree, is as big, says Dampier, and as long as a man's arm, and, all clustering around, grows the fruit—and such fruit! The Spaniards give it the first place among the productions of Pomona as most conducive to life. It grows in a pod about six or seven inches long, and is of the size of a man's arm—a favorite comparison, we may observe, with Dampier. The pod, shell or rind is soft, and when ripe is as yellow as gold. The fruit within is no harder than butter in winter, and resembles good yellow butter in color. It is of a delicate taste, and melts in one's mouth like marmalade. It is all pure pulp, without any seed, stone or kernel. Europeans when they settle in America learn to esteem it so highly that when they make a new plantation they usually begin with a good "plantain walk," as they call it, or a "field of plantains," and as their family increases, so do they enlarge their plantain walk, keeping one man purposely to prune the trees and gather the fruit as it reaches maturity. For some, or other of the trees, are always bearing throughout the year, and frequently

TEA PLANT OF JAPAN.

this is the only food on which a whole family exists. Such, at least, is Dampier's statement; but accurate as he generally is, some exaggeration is surely manifest here.

I have quoted thus far from the memoirs of Dampier, to show the style of writing, as well as the observing powers of the great buccaneer. Narrating in a period about two hundred years past, he writes on many subjects with a detail and accuracy truly astonishing.

In those days, as well at the present time, the plantain and banana are often confounded as one and the same fruit. In reality the distinction is nearly as great as that between a pumpkin and a melon. And although belonging to the same botanical species, one is a delicious natural fruit while the other requires the culinary art to make it acceptable as a food.

Manilla, on the island of Luzon, the capital city of the Phillippines, is in north latitude 14 deg. 36 min., and east longitude 120 deg. 52 min. Inhabited by about 300,000 people, it has long been the principal commercial port of the Spanish possessions in the Pacific. The exports of sugar are about 150,-000 tons per annum, with 50,000 tons of Manilla hemp and 100,000,000 cigars. In the manufacture of the latter, 10,000 women are employed, the factory covering a space of over six acres.

TOBACCO.

Nature, climate and locality have combined to make the islands of the Pacific favored lands for the extensive cultivation, preparation and export of tobacco.

Of the plant itself, it may be claimed as the *Nicotiana Tabacum*, indigenous to America, but cultivated now in nearly all parts of the world. Seeds of the plant were sent by Jean Nicot, in 1560, from Portugal, to Catherine de Medici. Nicot was the French ambassador in that country, from whom the plant receives its botanical name. Its first introduction into Europe from the new world may be dated from the beginning of the sixteenth century. Its first introduction into England by Sir Walter Raleigh, from Virginia, occurred about 1586. Hayden ascribes it to Sir John Hawkins in 1565, while many others grant it to Raleigh and Sir Francis Drake.

Baird, Humboldt, and many of the encyclopædias, state its name to be derived from the Indian word *tabacos*, a name given by the Carribees to the pipe, in which they smoked the leaves of the plant. Baird says it is the common name of the species of herbaceous, rarely-shrubby plants, of the genus *Nicotiana*, generally clothed with clammy hairs or down, and natives for the most part of the warmer portions of America, a few growing also in the East. The species which yields most of the tobacco of commerce, is the common Virginian or sweet-scented tobacco, extensively cultivated in the warmer portions of the United States.

The claim for its first uses among the Chinese, Mongols, and the East Indians, says Mr. McCulloch, is, however, a very doubtful proposition. It seems sufficiently established that the tobacco plant was first brought from Brazil to India about the year 1617, and it is most probable that it was thence carried to Siam, China, and other Eastern countries. The names given to it in all the languages of the East are obviously of European, or rather of American origin, a fact which seems completely to negative the idea of its being indigenous to the East.

Where properly cultivated, picked and cured, the best qualities of "Old Virginia" tobacco, for chewing or smoking, has no superior. That of Havana, for the manufacture of cigars alone, takes first place, but does not seem to have the requisite qualities that go to make either a palatable "fine cut" or "plug" chewing tobacco.

CHAPTER IX.

ISLANDS

And yonder by Nankin, behold!
  The tower of porcelain, strange and old,
Uplifting to the astonished skies
  Its nine-fold painted balconies,
With balustrades of twining leaves,
  And roofs of tile, beneath whose eaves
Hang porcelain bells that all the time
  Ring with a soft melodious chime;

LONGFELLOW (Keramos.)

ISLANDS OF THE CHINESE EMPIRE.

A BRIEF glance at some of the islands belonging to China may not prove uninteresting. They may be set down at about forty in number, with an area of 35,000 square miles and a population of 4,500,000. Hainan, Formosa, and the islands of the Chusan Archipelago, are the most important.

HAINAN.

Hainan, in the China Sea, between 18 deg. and 20 deg. north latitude, and between 108 deg. and 111 deg. east longitude, has an area of 12,000 square miles and a population of 1,500,000. It is but fifteen miles from the mainland of China, the inhabitants being principally people of that country. The interior is very mountainous, and is said to be a desolate, barren region. The shore country, however, is very fertile, and is cultivated with all the skill of the Chinese agriculturist. Unlike Formosa, there are many good harbors indenting its shores. The products of the land are similar to those already mentioned, ranging from the tropical to those of the more temperate climes.

FORMOSA.

Formosa, somewhat larger than Hainan, having an area of 15,000 square miles, lies between 21 deg. 58 min. and 25 deg. 15 min. north latitude, and east longitude 120 deg. and 122 deg.; is separated from the mainland by a channel nearly ninety miles in width. The inhabitants, some 2,500,000 in number, are of the Chinese and Malay types.

The island is of evident volcanic origin, many traces of former eruptions being found, but wholly inactive at present. Mountain ranges traverse the land, many of whose peaks are covered with perpetual snow. There are no good harbors, making commerce and navigation to and from Formosa, exceedingly dangerous. The lands, where cultivated, are very productive. Nearly all the fruits of the tropics are grown, with rice, coffee, sugar and tobacco as staples. The forests abound in camphor, cinnamon, ebony and other valuable trees.

Formosa was first made known to Europeans by some returning Spanish seamen who had lost their vessel on the island's rocky shores in 1582.

The fisheries of these two islands are of great value, as also those of the Chusan Archipelago. Immense quantities are taken, cleaned, dried and sold in the markets of China. This valuable interest is not confined alone to these islands, but is of great commercial importance in nearly all of the island groups described.

Like Australia, in the surrounding seas, as many as thirteen hundred species of fish are known.

JAPAN.

Dai Niphon, the Japanese Empire, we know of, through history and tradition, as far back as 680 B. C. The island empire is embraced between latitude 23 deg. and 50 deg. north, and longitude 122 deg. and 153 deg. east. Thousands of islands (the official number is stated to be 4,000), stretched over the Asiatic seas, make a landed area of about 250,000 square miles, inhabited by 34,000,000 people.

The island chains and clusters are divided into groups, the more important being named Kurile, Kiushiu, Niphon, Riukiu, Sado, Shikokiu, Yezo, Goto, Oki, Iki, Oshima, Awaji, Hirado, etc.; the most noted cities on which are Tokio (formerly Yedo), Kioto, Ozaka, Nagoya, Hiroshuma, Sagii, Kagoshuma, Kanagawa, Samoda, etc.

HISTORY.

Our first knowledge of Japan was through the celebrated Venetian traveler, Marco Polo, who visited the empire in the thirteenth century. At a more modern period we hear of them through the efforts of the Catholic missionaries, and again from the Dutch explorer, Kæmpfer. It remained, however, almost a *terra incognita* until 1854, when the United States, through the efforts of Commodore Perry, succeeded in making a commercial treaty that opened up the isolated empire to the trade of the world.

The islands of Japan were probably peopled by the Chinese in 1000 B. C.—many traces of whom are to be found in the language, manners, religion, customs and agriculture of the Japanese to-day. The art of navigation, also, was well understood by them for many centuries.

NAVIGATION.

As early as the sixteenth year of the reign of the Emperor Suizin, 81 B. C., merchant ships and ships of war are spoken of as being built in Japan.

In the early periods their vessels must have been greatly superior in form and build to those of the present day. In fact, they were and are mariners of no mean order, and through this circumstance alone, if we add the storms, favoring winds and the ocean currents of the Kuro Shiwo in the north, and the Peruvian currents in the south, the peopling of North and South America can be traced back to the Japanese and Chinese.

The disappearance of Japanese vessels off of their coast, with their crews, never to return, whether through accident or design, have become so frequent as to require an imperial decree to check it. Under the reign of Shogoon Irzemitsu, about 1639, an edict was issued commanding the destruction of all boats built on any foreign model, and forbade the building of vessels of any size or shape superior to that of the present junk. By the imperial decree of 1637, Japanese who had left their country and been abroad were not allowed to return, death being the penalty for traveling abroad, studying foreign languages, introducing foreign customs or believing in Christianity.

About this time all junks were ordered to be built with open sterns and large square rudders, unfit for ocean navigation, as it was hoped thereby to keep the people isolated within their own islands. Once forced from the coast by stress of weather, these rudders are soon washed away, when the vessels naturally fall off into the trough of the sea and roll their masts out. The number, of which no record exists which have thus suffered during the last nineteen centuries, must be very large, probably many thousand vessels.

(Brooks on Japanese Wrecks.)

TOPOGRAPHY.

The topograpical features of Japan must of a necessity vary a great deal. Being a country wholly composed of islands, large and small, the physical features of mountains, valleys, lakes and streams, have not that extent and grandeur of older and larger countries. The rivers for this reason are not long, broad or of very great depth, and therefore inland navigation is not much in vogue. However some of the mountain ranges are very prominent, notably the volcanic peak of Fugisan, with an altitude of 14,000 feet, in the regions of perpetual snow.

EARTHQUAKES.

Geologically, the position of most of the islands is of so uneasy a foundation that a popular tradition of the Japanese, locates their empire on the back of a huge catfish. To the uneasy and angry motions of this fish they attribute their numerous earthquakes—as many as eighty of these *temblors* sometimes visiting them in one day. They are frequent, and at times very disastrous, the danger from fire in their wooden cities often adding to the horrors. On this subject a recent writer says: Besides the outbursts of frequent volcanic eruptions, no country is more frequently visited by destructive earthquakes. Kæmpfer enumerates six active volcanic mountains. Earthquakes, he says, are so frequent that the natives regard them no more than Europeans do ordinary storms. In 1855 a succession of earthquakes took place and lasted forty days, causing the destruction the best portion of the city of Yeddo, and the death, it is alleged, of 200,000 of its inhabitants. In 1783 the eruption of a volcano on the island of Kiusui, accompanied by violent earthquakes, destroyed in a single province twenty-seven villages. Another volcanic eruption took place in the same island in 1793, accompanied by earthquakes, which continued from March to June, and caused, according to official returns, the death of 53,000 persons, with a proportional destruction of property. On the 23d of December, 1854, an earthquake occurred which was felt on the whole coast. Of the town of Simoda, only a few temples and private edifices that stood on elevated ground escaped destruction. The fine city of Osacka, on the southeastern side of Niphon, was completely destroyed, and the capital, Yeddo, did not escape without injury. On the 10th of November, 1855, an earthquake at Yeddo is said to have caused the destruction of 100,000 dwellings and fifty-four temples and the death of 30,000 persons.

(Homan's Cyclo. of Com. and Nav.)

METALLURGY.

The empire produces all the valuable minerals in abundance, as also a good bituminous coal, which they turn into coke and use extensively in working the metals.

They are the masters of many secret processes in mineralogy and metallurgy, and in the inlaying of metals, one on the other, much used in the ornamentation of their bronzes, mingling gold, copper and silver in the most delicate and intricate designs, have never been equalled in Europe or America. In modeling in wax, to receive the clayey covering afterwards, preparatory to casting the designs in bronze or other metals, they show a knowledge and skill seldom equalled.

FLORA.

The vegetable productions of Japan are for the most part common to temperate regions. Timber, however, is so scarce that no one is permitted to cut down a tree without permission from the magistrate, and only on condition of planting a young one in its stead. The most common forest trees are the fir and cedar, the latter growing to an immense size, being sometimes more than eighteen feet in diameter. In the northern portion of the empire two species of oak are found, which differ from those of Europe. The acorns of one kind are boiled and eaten, and are said to be both palatable and nutritious. The mulberry tree grows wild, and in abundance ; the varnish tree (*rhus vernix*) abounds in many districts. In the south the bamboo cane, though a tropical plant, is found either in the wild or cultivated state, and is much used in their manufactories. The camphor tree is of great value here, and lives to a great age. Siebold visited one which Kæmpfor described as having been seen by him 135 years before. It was healthy and covered with foliage, and had a circumference of fifty feet. The country people make the camphor from a decoction of the root and stems, cut into small pieces. Chestnut and walnut trees are both found. Among the fruit trees are the orange, lemon, fig, plum, apple, cherry and apricot.

(Homans.)

As agriculturists the Japanese are fully equal to the Chinese; in fact, using all the methods of irrigation, rotation of crops, the use of manures, so much in vogue in the older country. They are experts in the handling of the silk-worm, turning its cocoon into all the forms of valuable silk; while in the growth and cultivation of the tea-plant they are unsurpassed.

INHABITANTS.

They are far advanced in horticulture as well, and far ahead of other nations in their methods of urging on or retarding the growth of plants. Thus the Camellia Japonica may be seen from a very diminutive growth to a tree forty feet in height, while the pine, cedar, and fruit, are represented in trees of mature growth, from two inches in height up to the natural growth common in other countries.

The Japanese are bold and daring mariners, and the only race in these regions who pursue the whale. They make many voyages to Kamptchatka and the Aleutian isles, making light of heat and cold, or hardships of any kind. It has only been through the re-

straint placed upon them by the severest of laws, that
has prevented these people from being known to the
maritime world centuries ago, and taking their place
amongst the most enterprising and boldest of navigators.

The many bays and inlets indenting the island
shores, swarm with shoals of fish, and they, with the
lakes and rivers, are covered by aquatic birds, affording an easy living to the poorer classes. Pearls of
great value abound along the shores; while the shell,
much valued here, is worked up in a thousand ways
as ornaments and inlaid work.   .

The people are an active, vigorous race, and very
intelligent; and although shut up for so many centuries, isolated from the outer world, they are kind and
hospitable to strangers, carrying their courtesy and po-
liteness to the greatest extreme. Since the American
treaty in 1854 they have steadily improved in shipping
and manufactures, freely admitting all our arts of peace
and war to be introduced among them. At the present time, young Japanese of the better classes are to
be found traveling in all parts of the world or attend-
ing the colleges and academies of the most advanced
nations, diligently and intelligently seeking all that
may advance or benefit their native land.

TEA PLANT.

A brief description of the tea plant, so assiduously
and profitably cultivated by the Asiatic races, may
interest the general reader.

The tea plant (*Thea Sinensis*), in a wild state, is
a bushy shrub, often reaching to the dignity of a tree
in size and foliage. In the cultivated state, in China

and Japan, the plants are held back, being pruned down and not allowed to grow higher than three to five feet. Botanists of to-day rank it as *Cammellia Thea* genus, same as the *Cammellia Japonica;* also bearing a close resemblance to the *Cammellia Sasanqua*, introduced in Europe and America from China in 1811.

The plant resembles the japonica somewhat in its buds and flowers, the leaves differing in being longer, narrower and less shiny. It is an evergreen, and affords from three to four crops a year, the second picking being considered the best. The leaves are picked altogether by hand, when they are conveyed to drying floors, the green varieties being dried on copper plates over slow fires, which results (not, however, without the assistance of being rolled between the hands of the laborers) in the closely-curled form found in nearly all teas. It has been stated that the green variety owes its color to the chemical action of the copper on the leaves. This is erroneous, as the black varieties are picked from the same plant, and receive their color from being allowed to go through a slow fermenting process, which changes the leaves of the same plant from green to black. From the dry-houses the tea is packed in lead-lined cases, or put up in paper packages, as we see it in the markets of the world.

Tea was first discovered in China, growing in a wild state, in the eighth century. In the fourteenth year of the reign of the Emperor Te-Tsong, corresponding to the year 783 of our era, we find an impost levied on tea. Japanese writers state that the plant was first brought to their country from China in the ninth century.

Of Europeans, the Portuguese were probably the first to discover its uses—in 1517. An Englishman—

one of the officers of the East India Company—speaks of it in a letter to his company in 1615.

In the years 1870, '71, '72 and '73, the imports of tea into Great Britain were about *sixty thousand tons* per annum, valued at about $55,000,000.

Into the United States, in 1871, '72, '73 and '74, the imports of tea were about 27,500 tons per annum, of an annual value of about $20,000,000.

CAMPHOR TREE.

A valuable indigenous growth of Japan is the camphor tree, one of the laurel family (*Lauracea Camphora*). It is native to the soil of nearly all the islands of the Eastern Archipelago and the Asiatic coast. The tree grows to a large size, with beautiful, wide-spreading foliage, and bears a small fruit, not unlike in size and appearance to a black currant. The ordinary camphor of commerce is produced by steeping the twigs, roots, and other portions of the tree in water, and then, by heat, distilling the liquid over into condensors, where it deposits in small white crystals, when it is carefully dried and packed for shipment.

That of Borneo, Java, Sumatra, and some of the Molluccas, is taken from the tree in the form of a gum, which exudes from the limbs, body and roots, drying and crystallizing in masses, sometimes weighing ten or fifteen pounds. This quality is considered to be of great value by the Chinese and Japanese, who readily pay a hundred times more for it than for that produced by distillation.

The wood of the tree is of considerable importance, being worked up in many ways into glove boxes, trunks, chests, and as a veneer for all receptacles requiring protection from the inroads of the insect world.

GOVERNMENT—RELIGION.

The system of government of the Japanese Empire is that of an absolute monarchy. The power of the Mikado is absolute and unlimited in legislative, executive and judicial matters. The Great Council (Daijo-Kwan), in which the Emperor himself presides, is the supreme executive, as well as the highest legislative body. It is composed of a Prime Minister, two Junior or Vice Prime Ministers, and a number of Privy Councillors. The latter, as a rule, are either heads of the several executive departments or other important bodies. At present there exists no complete severance between the legislative and executive sections of the Government. The most important body in the Government is the Gen-Roin, or Senate, established in 1875. It deliberates on legislative matters, but its decisions are subject to confirmation by the Great or Cabinet Council, and sanction by the sovereign. The number of Senators is unlimited (thirty-seven in 1883); they are chosen from those who have rendered signal service to the state. Another body, the Sanji-in (Council of State), created in 1881, has the function of initiating and framing bills, and discussing matters transmitted by the executive-departments, subject to deliberations in the Senate. It also hears and decides cases relating to administration.

The religion of nearly the whole of the lower classes is Buddhism, which had 76,275 priests in 1881; Shintoism had 17,851 priests. Christianity is stated to be spreading among the people. School attendance has been made compulsory.

LADRONE ISLANDS.

Due east from the island of Luzon, between latitude 13 deg. 50 min. and 20 deg. 50 min. north, and longitude 145 deg. 50 min. and 147 deg. east, are the Ladrone or Marian group.

There are in all about twenty islands, of which Guajan is the largest, being about 90 miles in circumference. The area of the group is 1,300 square miles, with a population of 8,000.

Discovered by Magellan in 1521, they form a part of the Spanish possessions in the Pacific.

The products are similar to the many islands already described, with an abundance of water, and soil of great fertility.

North by east from the Ladrones are the Jardines group, and north of these, again, Anson's Archipelago. Still further east and south we come to the Nameless group, Volcano, La Mira, Halcyon, Wakes, Cornwallis, and many other islands dotting the great Pacific Sea. In longitude 162 deg. 60 min. west, and 2 deg. north latitude, there is Christmas Island; and north by west from that, and in the same group, we find America, Fanning, Palmyra, Prospect and Samarang Islands.

To the north, again, in latitude 15 deg. 45 min. north, and longitude 169 deg. west, are the Johnston Islands, two in number, and of considerable commercial importance, from the guano found there.

BONIN ISLANDS.

The Bonin group, between 26 deg. 30 min. and 27 deg. 44 min. north latitude, and 142 deg. and 145

BANCROFT LITH S.F.

VOLCANO OF MT. EREBUS, Victoria Land, South Polar Regions; Lat. 77° 32' S.; Lon. 167° E.
Discovered by Capt. James C. Ross, January 27, 1841. P. 249.

deg. east longitude, may be set down as containing seventy islands, with twenty or thirty rocks lying between. There is no definite data at hand giving the area and population of this group, though it would be safe to set the former at 500 square miles, and the latter at 1,000.

The formation is volcanic, the topography rocky and precipitous, with deep water close to shore. They have long been a resort for whalers in these regions, for wood and water supplies.

The islands, at one time, in the latter part of the seventeenth and the beginning of the eighteenth centuries, were used by the Japanese as penal colonies. Pell, Buckland and Stapleton are the largest and best-known islands.

Their products are unimportant at present. The group is claimed by Great Britain, being taken possession of by that power in 1826.

ANSON AND AUCKLAND ISLANDS.

There are many island groups, atolls and barren isles, hardly as yet of enough commercial importance to require special or particular description. Under this head is the Anson archipelago, lying west of the Hawaiian group; and although but a chain of small islets, with but few products, it would be hard, in this age of discovery and requirements, to predict their future.

The Auckland Islands, between latitude 50 deg. 24 min. and 51 deg. 4 min. south, and longitude 163 deg. 46 min. and 164 deg. 3 min. east, are of considerable importance. They are about twenty in number, several of them, like the island of Auckland, being fully 30 miles long by 15 miles wide. They are of volcanic origin, with an abundance of water and timber and fertile soil. Guano of a fine quality is said to be in quantity on some of the islets. Discovered in 1806, they remained for many years almost unknown and unoccupied, up to 1849, when they were granted by Great Britain to a corporation, who used them principally as a whaling station, but were finally abandoned in 1852. The northern portion of the group is sometimes known as the Enderby Islands. The whole group may contain an area of 1,000 square miles, with a population of 500.

CHAPTER X.

ISLANDS

An island salt and bare,
The haunt of seals, and orcs, and seamews
clang.

MILTON (*Paradise Lost*).

ALASKA AND THE ALEUTIAN ISLANDS.

THIS chain of islands, stretching from Alaska in a southeasterly direction to the shores of Kamptchatka, lying between 51 deg. and 56 deg. north latitude, and 163 deg. and 188 deg. west longitude, form almost a connecting link between North America and Asia.

They are about fifty in number, and comprise within their limits nearly 8,000 square miles. They at one time formed a portion of the possessions of Russia in America, and were, with Alaska, deeded to the United States by purchase in 1867.

Unimak and Ounalaska are the principal and largest of the four different groups. From climatic reasons, as well as their long distance from the civilized world, they are very thinly populated and with little or no agricultural cultivation. Water is very scarce, while there is hardly any growth of timber, they present a picture not at all inviting to future pop-

ulation. Some of the valleys are well fitted for grazing purposes, abounding with nutritious grasses, while
the surrounding waters of the sea teem with fish.
The whale and the seal make these latitudes at one
time of the year a favorite resort, and are taken in
great numbers. There are about 3,000 inhabitants
in the Aleutian group, whose existence must be anything but cheerful.

From their geographical situation, some writers
and ethnologists have supposed the Aleutian chain to
have formed the bridge between America and Asia,
over which the Asiatics crossed, gradually peopling
America.

The purchase price paid by the United States to
Russia for Alaska and the adjoining islands was $7,-
200,000. The late important developments being
made in that territory in minerals alone, gold, silver,
copper and coal, not to mention the immense forests
of valuable timber, leaves one with the impression
that our Government did a wise thing in its purchase.
Its area, something over three and one-half times that
of the State of California, for which we paid Mexico
$15,000,000, may yet prove it a veritable bonanza.
Probably not in an agricultural way, but in fisheries,
minerals and timber it may exceed all our past fortunate experiences in territorial acquisitions, like California, Arizona and New Mexico, etc.

ISLANDS OF ST. PAUL AND ST. GEORGE.

Two of the islands, St. Paul and St. George, have
been found to be the favorite resorts of the fur seal.
This was taken advantage of by a San Francisco corporation, who leased the Islands from the Govern-

ment at a yearly rental of $55,000, for the purpose of a seal fishery alone. They are restricted to taking but 100,000 a year, on which the United States receives a tax of $2.62½ each, producing in all a revenue to the Government from rental and tax of $317,500 per annum.

The island of St. Paul is located in north latitude 57 deg. 8 min., and west longitude 170 deg. 13 min. St. George lies about forty miles to the south. From the former, 80,000 seals are taken each year; from the latter, 20,000.

SEALS AND SEAL FISHING.

From "Dall's Alaska and its Resources," published in 1870, we learn that the fur seal fishery, formerly less important than that of the sea otter, has of late years far exceeded it in value. A short review of the history of this fishery may not be out of place. At present the fur seal are almost exclusively obtained on the islands of St. Paul and St. George in Behring Sea. A few stragglers only are obtained on the Falkland Islands and the extreme southwest coast of South America. The case was formerly very different. Many thousands were obtained from the South Pacific Islands and the coasts of Chili and South Africa.

The Falkland Island seal (*Artophoca Falklandica*) was at one time common in that group and the adjacent seas. The skins, worth fifteen Spanish dollars, according to Sir John Richardson, were from four to five feet long, covered with reddish down, over which stiff gray hair projected. They were hunted especially on the Falkland Islands, Terra del Fuego, New Georgia, South Shetland and the coast of Chili.

Three and a half millions of skins were **taken** from Mas a Fuera to Canton between 1793 and 1807. Another species (*Artocephalus Delandi*) formerly abounded on the coast of Africa, near the Cape of Good Hope. Their fur was the least valuable of the different kinds of fur seal, and the species seems to have become extinct. * * *

Of the Arctic or Behring Sea species (*Callorpinus Ursinus*) not less than 6,000,000 skins have been obtained since 1741.

HABITS.

The Alaskan fur seal formerly extended from the ice line of Behring Sea to the coast of Lower California. At present a few stragglers reach the Straits of Fuca, where 5,000 were said to have been killed in 1868, but the great majority are confined to the Pribyloff Islands. They have never been found in Behring Strait, or within 300 miles of it. They arrive at the islands about the middle of June, a few stragglers coming as early as the end of May. They leave on the approach of winter, usually about the end of October. They are supposed to spend the winter in the open sea, south of the Aleutian Islands.

When returning from their winter quarters (the location of which is yet unknown), they come up in droves of many thousands on the hillsides near the shore, and literally blacken the islands with their numbers. * * *

METHOD OF KILLING SEAL.

The manner of conducting the fishery is as follows: A number of Aleuts go along the water's edge, and getting between the animals and water, shout and

wave their sticks.   The seals are very timid, and always
follow each other like sheep; yet, if brought to bay,
they will fight bravely.   A man who should venture
into the midst of a herd would doubtless be torn to
pieces, for their teeth, though small, are exceedingly
sharp.

A body of four or five hundred having been separated, as above, from the main assembly, they can be
driven very slowly by two men into the interior of the
island, exactly as a shepherd would drive his sheep.
Their docility depends upon circumstances.   If the
sun is out and the grass dry, they cannot be driven
at all.   If the day is wet and the grass sufficiently
moi t, they may be driven several miles.   Every two
or three minutes they must be allowed to rest.   Those
who become tired are killed and skinned on the spot
by the drivers, as it is of no use to attempt to drive
them.   They would at once attack the driver. * * *
When the seals have been brought to a suitable place,
they are left with some one to watch them until it is
desired to kill them.   The skins of old males are so
thick as to be useless.   The Russians restricted the
killing solely to young males less than five years, and
more than one year, old.

No females, pups or old bulls were ever killed.
This was a necessary provision to prevent their extermination.   The seals are killed by a blow on the
back of the head with a heavy sharp-edged club.
This fractures the skull, which is very thin, and lays
them out stiff instantly.   The Aleut then plunges
his sharp knife into the heart, and with wonderful
dexterity, by a few sweeps of his long weapon, separates the skin from the blubber to which it is attached.
The nose and wrists are cut around, and the ears and

tail left attached to the skin. When the operation is
over the skin is of an oval shape, with four holes,
where the extremities protruded. They are then
taken out and laid in a large pile, with layers of salt
between them. After becoming thoroughly salted,
they are done up, two together in square bundles,
and tied up with twine. They are then packed for
transportation to London. No guns are used in killing the seal. Indeed, guns are not only unnecessary,
but injurious, for a hole in the skin diminishes its
value one-half. All the fur seals are dressed in
London. They were worth in the raw state, in 1868,
about $7 each in gold. (Now, 1884, said to be worth
$10 each.) A machine has been invented by which
the skin is shaved very thin, the roots of the stiff hairs
are cut off and they may then be brushed away.
The down, which does not penetrate the skin to any
distance, remains, and is dyed black or a rich brown.
This is the state in which we see the skins at the
furriers.

VANCOUVER ISLAND.

Vancouver Island, in and between latitude 48
deg. 18 min. and 50 deg. 55 min. north, and longitude
123 deg. 15 min. and 128 deg. 30 min. west, has an
area of about 14,000 square miles, with a population of
15,000.

The principal products are coal and timber of a
fine quality. Of the former, immense quantities are
produced.

Although my purpose throughout has been to re-
frain altogether from allusion or description of lands
not located strictly as ocean islands, yet so grand and
interesting is Puget Sound, that the following short
description may be of interest to the general reader.

THE PUGET SOUND REGION.

Puget Sound abundantly deserves its reputation for remarkable beauty. Commodore Wilkes is quite within the bounds of truth in his statement: "Nothing can exceed the beauty of these waters. I venture nothing in saying there is no country in the world that possesses waters equal to these." With a length of probably not more than two hundred miles, the sound has a coast line of more than fifteen hundred miles. It covers an area of about two thousand square miles, or a little more than twice the extent of Cook County, in which Chicago is. Its waters are very deep, and at almost any point vessels of the largest size may approach to land until their sides touch the shore, before their keels touch the bottom. It has hundreds of beautiful islands and bays. It lies as a deep basin between two lofty ranges of mountains —the Cascade Range on the east, and the Coast or Olympian Range on the west. The gateway opening into it from the Pacific Ocean is the Straits of Juan de Fuca (the name of their discoverer), which are ninety-five miles long and an average of eight miles in width. The sound itself was named for Peter Puget, one of Vancouver's lieutenants, who explored it. This great navigator gave to another of his lieutenants, Rainier, the honor of calling the grandest mountain peak in the country by his name, though it is now more generally called by the Indian name Tacoma (nourishing breast), while it is claimed that its true Indian name is Tanoma (almost to heaven). It is the highest peak but one in the United States, Mount Blanca in Colorado being just twenty feet higher. The latter, however, is not so massive, so grand, so overwhelming to the view,

since no beholder looks upon it except from an elevation of as much as seven thousand feet, while the former, at the town of New Tacoma, is seen from the sea
level, rising grandly 14,444 feet, and covered perpetu-
ally with snow and ice, its glaciers surpassing, in extent and grandeur, anything to be seen in the Alps.
Senator Edmunds, who visited the mountain last year,
says of it: "I have been through the Swiss mountains,
and I am compelled to own that—incredible as the
assertion may appear—there is absolutely no comparison between the finest effects that are exhibited there,
and what is seen in approaching this grand isolated
mountain. I would be willing to go five hundred miles again to see that scene. This continent is yet in ignorance of the existence of what
will be one of the grandest show places, as well as a
sanitarium."

QUEEN CHARLOTTE ISLANDS

Northwest of Vancouver, one hundred and thirty
miles, and eighty miles from the coast, are the Queen
Charlotte Islands. Like Vancouver, they belong to
British Columbia.

There are, in all, about twenty islands in the
group, the principal being Prevost, Graham, North
and Moresby.

Area of the group, 5,000 square miles; population, 6,000. The climate is good, with an abundance
of water, and pine and cedar timber. Copper, iron and
coal are found.

Many good harbors are to be met with in the
group, while the bays and inlets around the islands
teem with fish.

ISLANDS—WEST COAST UNITED STATES.

The Farralones consist of two clusters, comprising seven islands, the nearest of which is about twenty miles from the Golden Gate. They are all destitute of soil and vegetation, consisting of bare, rugged rocks, which are the resort of immense numbers of sea lions and myriads of birds, the eggs of which were a source of great profit to those who collected them.

The southernmost of the group is the largest, containing about two acres, and is also nearest to the the coast. On this there is a first-class lighthouse to warn the mariner of the dangers of the locality.

No water fit for drinking, except such as was collected from rains and fogs, was obtainable on any of of these islands until 1867, when some of the egg gatherers discovered a spring on the main island, near the lighthouse.

There are no other islands on the coast of California north of Point Concepcion. South of that headland there are two groups, the most northerly consisting of the islands of San Miguel on the west, Santa Rosa in the center, and Santa Cruz on the east.

Santa Cruz, the largest of this group, is twenty-one miles in length and four miles wide, and has a rugged surface.

Santa Rosa is fifteen miles in length and nearly ten miles wide. Its surface is tolerably level, and produces a thick crop of coarse grass and low bushes, but its steep, rugged sides, which rise nearly two hundred feet, almost perpendicularly, afford no good landing place.

San Miguel is nearly eight miles long and from two to three miles wide. It is almost a barren rock, but several thousand sheep manage to subsist upon the limited pasturage growing on the island. About forty miles southeast from the above cluster, and off the coast opposite Los Angeles County, are the islands of San Nicolas and Santa Barbara, and still farther in the same direction are Santa Catalina and San Clemente.

San Nicolas, the most western, is nearly sixty miles from the main land. It is eight miles in length by about four wide. Its surface is a flat ridge, nearly 600 feet high, tapering down in rocky ledges to the sea.

Santa Barbara Island is nearly circular in outline, and about two miles in diameter at the base, its surface on the top containing about thirty acres.

Santa Catalina, the largest island of this group, is about 400 miles south from San Francisco and twenty-five miles from San Pedro, its nearest point to the main land. It is nearly twenty-eight miles in length, about seven miles wide on its southern and two miles on its northern end. Its surface is rough and uneven, some points being 3,000 feet above the sea level; but it contains several small valleys which are under cultivation. * * There is a small stream of water running through its entire length; it also has a number of springs of fresh water. ֎ ֎

San Clemente, the most southern, lies about fifty miles from the main land of San Diego county. It is twenty-two miles in length by about two miles wide. * * It contains neither soil, vegetation or water. * *

(Cronise, Natural Wealth of California.) .

PACIFIC ISLANDS OF MEXICO.

Of the islands off the coast of Lower California, and in the Gulf of California, belonging to Mexico, there is little to be said.

In the Gulf, Carmen and Tiburon are the largest and most important. The former has long been celebrated for the immese quantities of salt exported, while of the latter but little is known, a hostile tribe of Indians being in possession. .

On Carmen, several hundred yards back from the seashore, nature has placed a salt lake, probably one-half a mile in diameter, a great natural evaporating pan, which furnishes a continuous supply of salt, that covers its surface like a crust of glistening snow. This is raked together in snowy heaps and taken away on hand-cars, running on several tramways built out into the lake. So rapid is the evaporation and accumulation of the salt that hardly the length of a day transpires before another supply is ready for removal. This salt marsh has been in operation for over twenty years, and the supply is undiminished.

Off the coast of Lower California the islands of Guadalupe, Cerros, San Benito, Lobos and Santa Margarita are of some size and importance. Now but the homes of innumerable wild goats, the day may come when the finer breeds of the Angora will be introduced, and make these barren spots the source of valuable industries.

Further south, the island groups of Tres Marias, Revilliagigedo, etc., are to be met with, and although not of great extent, are of considerable value from the pearl and other fishing grounds found there. The fine timber of the tomano and prima vera, much used in the manufacture of furniture and cabinet ware on the Pacific coast, is exported in large quantities.

The pearl fisheries of the Gulf of California and the Bay of Panama form quite an industry, the pearls and shell found often being of the best quality. Pearl, the shell, and fisheries, have been noticed at some length in another portion of this work, although some of the suggestions made in the chapters on that subject might be applied in these localities with great profit.

CHAPTER XI.

ISLANDS

A wilderness of sweets.
MILTON (*Paradise Lost*).

THE SANDWICH ISLANDS.

THE Hawaiian group, 10 in number, although some writers say there are thirteen, is between latitude 18 deg. 54 min. and 23 deg. 34 min. north, and 154 deg. 50 min. and 164 deg. 32 min. west longitudes. The total area is near 6,000 square miles and the population some 65,000.

The rapid growth of this little island kingdom, and that within a very few years, into commercial importance, is but a sample of what will be done in the *island world* in the near future. The topographical features of the group, the few and small streams, with valleys of no very great extent, with a wasteful destruction of nearly all the valuable indigenous products in the past, with the low order of inhabitants, has barred their progress, yet the magic wand of American enterprise has but touched them, and the islands are now in practical, successful commercial existence.

The principal export is sugar. Of this valuable

product it is safe to say that 150,000,000 pounds, or 75,000 tons will be produced this year.

GEOLOGY.

The geological formation of the group is altogether volcanic. Two celebrated volcanoes, Kilauea and Mauna Loa, are noted for their eruptions, and in some of their convulsions the world-famed outbursts of Ætna and Vesuvius "pale their ineffectual fires." Thus, in the island of Hawaii, according to the Journal, Geological Society, 1856, in 1840 a deluge of lava broke out ten miles below the crater of Kilauea. It spread from one to four miles wide, and reached the sea at a distance of thirty miles in three days, and for fourteen days plunged in a vast fiery cataract a mile wide over a precipice of 500 feet. In 1843 a similar stream flowed from the summit of Mauna Loa, and in 1855 the lava broke out at a spot 2,000 feet below the summit, on the opposite side to Kilauea, and continued for ten months, overflowing an area of 200,000 acres. · The main stream was sixty-five miles long, from one to ten miles wide and from ten to 300 feet in depth. The records do not show any eruptions of Mauna Loa previous to 1832. There were outbursts in 1851, '52, '55 and '59. In 1868 one occurred accompanied by a severe earthquake. The last was in 1877.

SUGAR CANE.

The wonderful sugar-producing qualities of this little island group, now something like 70,000 tons per annum, is gradually calling the attention of the world to what might be done on other islands of the Pacific.

THE KING OF ST. GEORGE—ALASKA.

Many of these garden spots are peculiarly adapted to the growth of cane; the soil, climate and moisture necessary to its successful cultivation being found on every hand.

The sugar cane, *saccharum officinarum*, is a perennial plant, of the family of grasses, cultivated sorghum and broom-corn being familiar examples of the species. The cane is not found native in any country, never producing seeds, and is only reproduced from cuttings. There are many varieties, but the best is the Otaheite, or Bourbon, grown successfully in the Society group.

Sugar is mentioned at a very early period, being used then only as a medicine. It was introduced into Persia about the ninth century. In the tenth century it was cultivated and formed an article of trade in Spain.

It was first cultivated in Madeira in 1420, and shortly afterwards in the Canary Islands. After the discovery of America it was introduced into Mexico, San Domingo, Brazil, etc., and about the same time into Africa and the Indian Archipelago. In our own country it was first cultivated by the Jesuits, near New Orleans, in 1751.

HISTORY.

In regard to the discovery of these islands by Captain Cook, I am led to believe that he was by no means the original discoverer, but that like many other navigators on the great oceans of the world, it was a discovery for him, while in truth it may have been known to others many ages previous.

It is believed that the Hawaiian Islands were first discovered by the Spaniards, and were often seen by the Spanish galleons on their yearly passages between Acapulco and Manilla in the sixteenth century. According to tradition, two Spanish vessels from Mexico were wrecked on the island of Hawaii about the year 1525. Their crews mixed with the native race, whose descendants, it is said, can even now be distinguished by their complexion.

The Spanish charts of the Pacific Ocean, dated in the sixteenth century, give the position of the islands with some accuracy, and call them by names, describing the appearance which each island presented to the Spanish navigators when seen from their vessels. These charts were known in England when Captain Cook sailed on his voyages of discovery; and as the London charts of 1777, the year before Cook first visited the islands, record their existence, this English navigator cannot be considered as their discoverer.

About the year 1740, according to tradition, a ship landed a crew of white men on the island of Oahu. The natives knew the value and uses of iron before Cook arrived. They stole his boat and broke it up to get the iron from it, in Kealeakua Bay, where his ships anchored in January, 1779, and where he was killed in a combat with the natives on the 14th of February, while negotiating, on the shore, for the return of his boat.

The French navigator, La Perouse, who also was killed by Pacific savages, visited the islands in 1786. In 1790 the first trading ship arrived—the American ship *Eleanor*. The English explorer, Vancouver, arrived in 1792, and brought from California the first cattle that the islands had seen. In 1793 the harbor of Honolulu was discovered and entered by a trading vessel from the west coast of America. In 1820 the first whaler arrived—the ship *Mary*, from Nantucket. The lighthouse of Honolulu was first lighted in 1869.

The first Protestant missionaries arrived at the islands in 1820, by the brig *Thaddeus*, which sailed from Boston in 1819. * * *

They were well received by the islanders, who were superstitious idolaters, living under the tyranny of their chiefs and priests. Since 1820 the American churches (up to 1873) have sent nearly one hundred and fifty men and women as missionaries to these islands, and have spent a million of dollars for their evangelical civilization.

One result of the investment is the controlling influence of Americans, etc.

(Bliss's " Paradise in the Pacific.")

COTTON.

The cotton plant (genus *gossypium*) is an indigenous growth of nearly all the intertropical countries, there being as many as eight varieties of the plant— one (the *gossypium sandwichsense*) being native in the Sandwich Islands.

In India, cotton, its cultivation and uses, have been known since prehistoric times, and was introduced from there into Japan in the seventh century, and into Europe by the Mohammedans in the tenth century. In the United States it was known as far back as 1536 —the product from the latter country being now about one and one-half million tons per annum. The finer qualities—that with the longest fibre—grown in the United States, on the isles along the coast of Georgia and some of the other seaboard States, known as sea-island cotton, would naturally suggest it as one of the staples to be cultivated in the South Sea.

Hawaii, although the largest island in the group (having an area of nearly four thousand square miles), has but one harbor of any note—that of Hilo. There are many indentations along the shores that might serve as good anchoring for vessels, but the sterility of the back country has so far prevented their occupation and settlement.

Next in size is Mau, with about 603 square miles of area. There is little to be said, except that the lands are extremely productive where placed under cultivation.

On Oahu, third in size, whose area is 522 square miles, is located the principal port of entry and harbor, as well as the capital city of the kingdom, Honolulu. There is good anchorage here, with a barrier reef of coral and lava protecting it on every side.

Deep-water soundings are found on every hand before entering, while inside the average depth may be set down at about twenty feet.

The volcanoes of Mauna Loa and Kilauea, located on the island of Hawaii, have already received some attention in this work.

Kilauea is sometimes claimed to have the largest active volcanic crater in the world, having a circumference of over eight miles, and a depth, from the rim of the basin to the burning lava, of one thousand feet. The elevation of the crater is 4,000 feet, while its fiery neighbor, Mauna Loa (both being located on the mountain of that name) towers into the skies 13,760 feet above the sea. In regard to the size of volcanic craters, it might be said that in the eastern part of the island of Java a crater is to be found twelve to fifteen miles in circumference; that of Kilauea does not exceed nine miles. On the eastern peninsula of the island of Maui, one of the Sandwich group, is located the summit crater of Mauna Haleakala, 10,200 feet above the sea level. Following the rim of the once fiery cauldron, the circumference is all of twenty-seven miles, while the depth from the edge to the bottom of the great pit is two thousand feet. As far as known, this is the largest volcanic crater in the world. Of Kilauea, Dana says :

BURNING LAKE OF KILAUEA.

Kilauea is a deep pit in the sides of Mount Loa. The gentle slopes of the dome in this part scarcely vary from a plain, and the crater appears like a vast gulf excavated out of the rock-built structure. Although there is no cone, the country around is slightly raised above the general level, as if by former eruptions over the surface; but this is hardly apparent without extended and careful examination.

The traveler perceives his approach to the crater in a few small clouds of steam rising from fissures not far from his path. While gazing for a second indication, he stands unexpectedly upon the brink of the pit. A vast amphitheatre, seven miles and a half in circuit, has opened to view. Beneath a gray, rocky precipice of 650 feet, forming the bold contour, a narrow plain of hardened lava (the "black ledge") extends like a vast gallery around the whole interior. Within this gallery, below another similar precipice of 340 feet, lies the bottom, a wide plain of bare rock more than two miles in length.

The eye naturally ranged over the whole area for something like volcanic action, as it is usually described. But all was singularly quiet. In the dark plain that forms the bottom there was little to attract attention beside the utter dreariness of the place, excepting certain spots of a blood-red color, which appeared to be in constant yet gentle agitation. Instead of beholding a sea of molten lava "rolling to and fro its fiery surge and flaming billows," we were surprised at the stillness of the scene. The incessant motion in the blood-red pools was like that of a cauldron in constant ebullition. The lava in each boiled with such activity as to cause a rapid play of jets over its surface. One pool, the largest of the three then in action, was afterwards ascertained by survey to measure 1,500 feet in one diameter, and 1,000 in the other; and this whole area—into which the capitol grounds at Washington might be sunk entire—was boiling, as seemed from above, with nearly the mobility of water. Still all went on quietly. Not a whisper was heard from the fires below. While vapors rose in fleecy wreaths from the pools and numerous fissures, and above the large lake they collected into a broad canopy of clouds, not unlike the snowy heaps or cumuli that lie near the horizon on a clear day, though changing more rapidly their fanciful shapes. On descending afterwards to the black ledge at the verge of the lower pit, a half-smothered, gurgling sound was all that could be heard from the pools of lava. Occasionally there was a report like that of musketry, which died away and left the same murmuring sound—the stifled mutterings of a boiling fluid. Such was the appearance of Pele's pit in a day view, at the time it was visited by the author (in November, 1840).

At night, though less quiet, the scene was one of indescribable sublimity. We were encamped on the edge of the crater, with the fires in full view. The large cauldron, in place of its bloody glare, now glowed with intense brilliancy, and the surface sparkled with shifting points of dazzling light occasioned by the jets in constant display. A row of small basins on the southeast side of the lake were also jetting out their glowing lavas. Two other pools in another part of the pit tossed up their molten rock much like the larger cauldron, and occasionally burst out with jets forty or fifty feet in height. The broad canopy of clouds above the pit, which seemed to rest on a column of wreaths and curling heaps of lighted vapor, and the amphitheatre of rocks around the lower depths, were brightly illuminated from the boiling lavas, while a lurid red tinged the distant parts of the inclosing walls, and threw into shades of darkness the many cavernous recesses. And over this scene of restless fires and fiery vapors, the heavens, by contrast, seemed unnaturally black, with only here and there a star like a dim point of light. The next night streams of lava boiled over from the lake, and formed several glowing lines diverging over the bottom of the crater. Towards morning there was a dense mist, and the whole atmosphere seemed on fire. Through the haze the lakes were barely distinguished by the spangles on the surface that were brightening and disappearing with incessant change.

ISLAND FORMATION.

Among the groups of Polynesia the Hawaiian exceeds all others in geological interest. The agency

of both fire and water in the formation of rocks is
exemplified not only by results, but also by processes
now in action, and the student of nature may watch the
steps through the successive changes. He may descend to the boiling pit and witness the operations in
the vast laboratory with the same deliberation as he
would examine the crucible in a chemist's furnace. Thus
the manner in which mountains are made and islands
built up becomes a matter of observation. The volcanic dome may be seen in process of accumulation
from overflowing lavas, and may be traced as it increases in size. Again, disruptions of the accumu-
lated rock may be observed, followed by their disappearance in the lavas below.

While these volcanic mountains are still extending
their limits in one part of the group, in others those
changes are finely illustrated, which they undergo
through the action of water, gradual decomposition
and other allied causes, and these effects are in every
stage of progress. In some instances the slopes retain
the even surface of the lava stream ; in others, they are
altered in every feature—the heights are worn down,
the whole surface gorged out with valleys, and the
depth of these furrowings of time, indicate that the
several islands differ widely in the length of the period
since they were finished by the fires and left to the
action of the elements.

Moreover, the coral formations of the shores present us with reefs now in progress from the growing
zoophytes, and there are also reefs elevated many
feet above the sea, having a close resemblance to beds
of limestone. Besides these, there are hills of drift sand-
rock, of coral origin. The various facts illustrate, therefore, all the results of coral growth and accumulation.

The group is consequently the key to Polynesian geology. It combines all the features which are elsewhere widely scattered, and they are so exhibited in progressive stages as to afford mutual illustration. An island like Tahiti, so broken into peaks and ridges, may excite wonder and doubt. The Hawaiian group suggests the same difficult problem as Tahiti, but an intelligent solution is at the same time presented for our contemplation and study.

(Dana, Geol., Wilkes' Exp. Expedition.)

ISLANDS PACIFIC COAST OF SOUTH AMERICA.

Off the west coast of South America there are at least 300 islands, becoming more numerous and in larger groups as we go towards Cape Horn.

. Those off the coast of Ecuador, the Albemarle, James. Chatham, Indefatigable, Hood, Charles, Narboro, etc., have already been alluded to in this work, under the head of Galapagos.

GUANO.

Lying near the coast of Peru, and only about twelve miles from the main land, between 13 deg. and 14 deg. south latitude, and containing but a few square miles of area, are the celebrated guano group, the Chincha Islands. It may not be unininteresting to state ·here, that nearly 20,000,000 tons gauno have been exported to Europe and America from this little group alone. The shipments were commenced in 1841, and continued on a scale of great magnitude up to 1872, when the guano deposits were practically exhausted. Between the years 1853 and 1872.

8,000,000 tons were shipped. It is said that the government of Peru was enriched from this source alone. If we admit Peru as having received $5 per ton for these deposits, it will be seen that bonanzas do not always lie in mineral veins.

As a fertilizer for the agriculturist, guano has no superior—one ton of it being equal to fifteen to thirty-four tons of the ordinary manures now in use.

There is no doubt but many islands of this character will ultimately be found scattered over the broad expanse of the South Seas. As guano is worth from $30 to $40 per ton in Europe and America, it does not require a great deal of figuring to show that any country or company making a discovery and location of this kind, will not only enrich themselves, but benefit the world at large. The islands of Ferrol, Guanape, Lobos, Tierra, Mengon, Pachacama, San Lorenzo and Zorati, also belonging to Peru, are of some importance.

The larger islands off the west coast of Chili are of great value, not only as important fishing grounds, but for the many agricultural products, and fine timber they produce.

The principal are Byron, Cambridge, Campana, Chiloe, Clarence, Desolation, Duke of York, Guaytecas, Hanover, Huafo, Landfall, Madre de Dios, Mocha, Narborough, Noir, Queen Adelaide, Santa Inez, Skyring and Wellington.

Chiloe is probably the most important, as well as one of the largest of the group, having an area of 5,200 square miles, and inhabited by some 10,000 people. It was first discovered by Mendoza, in 1588. Great attention is paid to agriculture; wheat, corn and potatoes being the favorite crops. With an abundant rainfall, and lands not too mountainous or hilly, Chiloe has long proved a source of wealth to her people. Many vessels, whalers and others, resort to these islands for their supplies, while from many of the islets lying between, considerable quantities of guano are shipped.

Some of the isles were at a former period favorite resorts for the fur seal, but like the islands of Juan Fernandez and Mas a Fuera, which were also great sealing grounds in their day, they have been driven away, and now make their breeding resorts on other groups.

EASTER ISLAND.

Due west from the northern coast of Chili, something like 2,300 miles, lies this little dot in the Southern Sea. It is located in south latitude 27 deg. 6 min., and west longitude 109 deg. 17 min., contains an area of about seventy square miles, and a population of 1,000 people, of the Polynesian type. Its discovery is sometimes credited to Captain Cook, in 1774, who visited it in that year; by others, to Roggewein, the Dutch navigator, who located and mentions it as early as 1722.

The island is of evident volcanic origin, three prominent craters of past eruptions being already discovered. The soil in the valleys, and some portions fringing the sea shore, is very fertile where placed under cultivation. There is but little forest growth, and water is scarce.

Of late years the island has assumed quite a prominence, from the remarkable features and evidences of a prehistoric race found there, to the great delight of scientists and the sunken continent theorists.

Hundreds of statues and broken columns are said to
be scattered over the land, some of the former being
of the human figure, fully forty feet in height, and
eight to ten feet broad across the shoulders.   Many
have fallen down, and others are rapidly succumbing
to the abrading influences of the elements, while others
again are found located in the volcanic craters themselves, and thought to indicate the ancient race, as fire
worshippers.   The rude sculpturing is from the common rock found on the island, many unfinished tablets
and statues being discovered in the quarries, as if the
inhabitants had been rudely interrupted in their work
by some awful volcanic outburst, or earthquake convulsion.

CHAPTER XII.

ISLAND PRODUCTS AND RESOURCES.

A pearl may in a toad's head dwell,
And may be found too in an oyster shell.

BUNYAN (*Apology for his Book.*)

PEARLS AND PEARL FISHING.

PEARL fishing has been a curious and valuable industry for ages, reaching away back into dim antiquity. The great demand of the present day, not only for the pearls, but for the mother-of-pearl shell, has made the industry a more valuable one than diamond mining. The innumerable uses that the shell is put to for ornament and for useful purposes, has created a continually increasing demand for it in all parts of the world. Among the islands of the Pacific, fisheries are found of vast extent, producing pearls and shell of the finest quality. In fact, some of these beds have furnished already bushels of the gem, ranging in value from one to thirty thousand dollars apiece; while shell, when properly cured and cared for, meets with ready sale in the principal markets of the world at about five hundred dollars per ton.

HABITS.

The pearl oyster has habits peculiar to itself; and as far as the writer has observed, all effort to change them or make any improvement in their condition or locality, has never been effected by man. All attempts to propagate or transplant the oyster from the localities where first found, have proved a signal failure. They are born and live and die, at or near their homes, and are not found hunting for fields or pastures new, or very far from the place of their nativity. It is a mistake, however, to suppose that the oyster does not or cannot move. The fallacy that they attach themselves in strings and clusters to the coral caves, in, under and beyond the surf in favorite localities, and never move from them, has been proven to the contrary. Places that have been cleared of the oyster altogether, by fishing, have been known, particularly after a great storm, to become thickly settled with new shell, and that, too, of a large size and apparent full growth; proving that they can swim, float, and move around as their needs and habits dictate. Their favorite breeding ground in the South Sea, and this only in particular localities, seems to be in and beyond the surf of some of the atolls, or horse-shoe shaped islands, that have a great lagoon in the center, and to and from which, the tide ebbs and flows without hindrance. In such places the small shell, from the size of a pea to that of a shilling, may be seen in great numbers, tossed about in the surf and on the waves; and again making their way with the inflowing tide to the lagoons of the atolls; there to sink to the bottom and form beds similar to those of the oyster of our own country. On the outside reefs and in deep water, say about twenty

fathoms, that being the greatest depth reached by the
native divers, the shell is of large size, sometimes
twelve to eighteen inches in diameter, and when
opened out measure from two to three feet across.
Generally speaking they are of no value except as
curiosities, never containing pearls, and have not that
beautiful prismatic coloring found in the regular shell.

PEARL DREDGING.

Pearl fishing, as practiced in different parts of the
world now-a-days, is rather a precarious business, being accompanied by great danger and many hardships. The poor divers soon wear out, and the slow
accumulation of shell, with here and there a pearl of
great value, makes the product worth all it will bring
in the market. It is not probable that much impetus
or safety could be given to the business by the general
use of submarine armor either, as it has yet to be managed by hand, and therefore slow progress is made.
The many thousands of square miles of pearl grounds
to be found among the islands of the Pacific, a great
deal of it as yet untouched, should suggest a more
rapid and effective mode of fishing. With this idea in
view I have consulted many of our practical mechanics
and engineers, who are engaged in building dredging
machines, as also those who manage them in their
practical workings in our rivers and bays, as well as
on the line of the canal now being cut through the
isthmus at Panama. From authorities like these I have
confirmed the idea that it is perfectly practicable to
handle nearly all the pearl fisheries of the Pacific
islands by means of steam dredgers. By such a
method vast quantities of shell could be brought to

the surface from depths not yet reached by the divers, and be opened, cleaned and assorted with a celerity that would no doubt astonish the natives.

In the interior lagoons, fishing in this manner would seem an easy matter, as the water is always smooth, not being affected even by the great storms sometimes experienced in these latitudes.

After reaching the age of seven or eight years, the pearl oyster appears to sicken and die, when it opens and spills whatever is contained in the shells. This being the case, it is a natural query as to what becomes of the pearls. They are never brought up by the divers, who are only seeking for perfect shell, and with the limited time they are under water—seldom exceeding two minutes—they break off and gather such as can be easily reached, and are glad to come to the surface for a breathing spell. If the divers, with or without armor, were employed only as prospectors, to locate the oyster banks, and steam dredgers brought into play for the effective work, there is no doubt that the business could be made immensely profitable.

PEARL DIVING.

Pearl fishing has not as yet been brought to a system, among the Pacific isles, commensurate with its value. True, the business has been prosecuted to a great extent at the Paumatou group, but hundreds of other favorable localities in these seas have hardly been prospected, and many are unknown.

At the island of Ceylon, under the encouragement of the English Government, the Cingalese have become experts in the business, although using nothing in the way of machinery to assist in its prosecution.

MARQUESAS ISLANDERS.

When diving in deep water, the Cingalese use a split stick or piece of bamboo clasped over the nose, and stuff their ears with wax or cotton, which leaves both hands free to gather and break off the shell, when found hanging to the coral and attached to the rock. (In the South Seas the business is conducted in an informal way.) In their fishing canoes there are generally four persons—two to manage and guide the boat and to assist the divers. Of the latter, two form the balance of the crew and dive alternately, thus giving each an opportunity to rest. When descending into deep water, a heavy stone is generally used, attached to one foot by a loose strap, and with sack and stone attached to a small line, which is paid out or hauled in, as may be required, by the assistants in the boat. Where shell are at all plentiful, the sack is soon filled, the foot slipped from the strap around the stone, a jerk given as a signal to the line, and the diver comes to the surface like a cork, while the weight and sack of shell are hauled up at their leisure. During the breathing spell, requiring at least thirty minutes, the other diver is making his preparations, and goes through the same process. When one or two hundred of the pearl oysters have been collected, they are carefully opened by means of a blunt, pointed knife, great care being exercised to preserve the edges of the shell, as well as a careful inspection of the oyster and its covering for any inlying pearls. These are sometimes found imbedded in the oyster itself, but, generally speaking, lie loose in the shell. One hundred pearls are often found in a shell, but are mostly small and of little value. They are pierced and put on strings, like beads, and used in nearly all countries as ornaments, known to the trade as seed pearls. The

12

divers do not go down in deep water, or to a depth of
forty to eighty feet, over ten or twelve times a day,
the strain upon the brain and lungs fatiguing them to
the last degree.   When through, the boats are pulled
for the shore, the empty shell piled carefully away in
the shade, so as to dry slowly—it being found that this
method preserves all the beautiful coloring of the
mother-of-pearl, and brings a much higher price in the
market.   These slow accumulations of pearl and shell
are kept up during fine weather, and at times when
the temperature of air and water are nearly alike.
When not already contracted for, as is the case in
nearly all pearl fisheries, the products are kept to
await the advent of some trading vessel, or (as has
unfortunately been the case among many of the groups
of islands) await the descent of some bold and ruthless
buccaneer.

NOTED FISHERIES AND GEMS.

Pearl fishing is conducted on a much more formal
scale at Ceylon, and opposite on the Cordatchy shore,
at the Sooloos and Bahrein Islands, and in the Persian
Gulf.   Boats are regularly employed at these places,
of ten or fifteen tons burden, with greater numbers in
the crews than I have mentioned.   On the Condatchy
shore, pearl fishing has been regularly followed as a
business for over two thousand years.

The price of pearls has changed very much in
modern times, probably from changes in manners and
fashions, and the admirable imitations that can be obtained at a low price.   One of the most famous pearls
was bought at Catifa in Arabia, by Tavernier, for the
fabulous sum of $550,000.   The one said to have been
dissolved and drank by Cleopatra was valued at

$403,645. Another of similar size and value was cut in two parts and used as ear-rings on the statue of Venus in the Pantheon at Rome.

Probably the largest pearl ever found belonged to the late Mr. Hope. It measured two inches in length, and had a circumference of four inches, weighing 1,800 grains.

One found in the Paumotous, South Sea, was sold to Queen Victoria for $30,000.

That of Sir Thomas Gresham, ground up and eaten at a banquet given to the Spanish ambassador, added a value to the dinner of some $45,000.

It is estimated that the Paumotou group of the South Sea has already furnished over thirty thousand tons of merchantable shell, and some millions of dollars worth of pearls, to the world.

PROPOGATION.

The subject of propogating and cultivating the pearl oyster has received a good deal of scientific investigation for many years, but up to the present has met with but little success. The following, taken from a recent publication, would indicate that a problem that has puzzled the world for ages is about to be solved:

Some time back the French Government sent the Secretary of the College of France to Tahiti to study the best means of preventing the exhausting of the pearl oyster beds at Papeet. The results of his experiment tend to show that this jewel bearing bivalve may be cultivated in a way similar to that practiced in the case of its edible relative. Like the European and American oyster, the pearl oyster is pronounced by Mr. Bouchon Brandely to be uni

sexual and to be physiologically constructed in the
same manner. The Mollusk Polynesia has also the
power of re-attaching itself to the coral reef when
cast back into the sea by the coral fishers in the event
of the smallness of its size proving it to be worth-
less. The experiments also tend to show that the
pearl oyster found around the coasts of the French
islands of Oceanica will thrive just as well in parks
and beds only six or eight feet beneath the surface, as
in the deepest water. As the oyster is unisexual it is
easy to produce artificial fecundation. All that is necessary is to bring the male spawn (the milk-like fluid found
in an oyster at certain seasons of the year) in contact
with the female spawn; a glass of sea-water will
suffice for this operat on. It is easy to distinguish
the one from the other, as the spawn, when mixed
with a little water, shows a granular formation perfectly visible to the naked eye, while the male spawn
retains its milk-like appearance. A few days after
the fecundation, appreciable results show themselves in
the form of microscopic oysters; the water must be
frequently removed, and as soon as the bivalves have
attained to a visible size they can be placed in parks,
where they are to remain until fully developed. The
results, considering the smallness of the expense involved, of establishing artificial beds of the pearl-
bearing oyster in French Polynesia, would be of in-
estimable value, for not only does this bivalve yield
the gem so highly prized by the ladies, but also the
mother-of-pearl, which the industrial art has a thous-
and and one ways to utilize. The pearl is in reality
mother-of-pearl, produced under special circumstances.
If the shell of the mollusk shall be pierced, or should
a grain of sand or other foreign substance find its

way into it, a growth of mother-of-pearl is formed either to stop the hole in the shell or to protect the delicate flesh of the mollusk from contact with the foreign substance. There is almost always a speck of sand or something of that sort in the center of a pearl, and the shells which contain large excrescences of mother-of-pearl usually show outward marks of damage. The oyster may even be forced to produce plants and mother-of-pearl by introducing some foreign substance into it, or by piercing the shell in such a way as to lay bare the flesh, but great care must be taken to do this without in any way injuring the bivalve. A great regularity of form, a brilliant white color with reflections similar to those of the opal, and size, are qualities that give the pearl its value. They are apt, however, to suddenly lose their brilliancy, but this evil is not without a cure, for if a pigeon is made to swallow such a damaged pearl, and killed within a few hours afterward, it will be found in its stomach, restored to all its original luster, a result due to the action of the gastric juice of the fowl and the intestines. Care must be taken not to leave the pearl too long; in the space of twenty-four hours it will lose one-third of its weight.

WHALE FISHERIES.

For a long period, many years before the revolutionary war, our people were noted for their push and enterprise in whale fisheries. No nation has been able to compete with them in a prosecution of a business that has simply become stupendous. As early as 1774, Burke, in his great speech on American affairs, paid a high compliment to the energy and enterprise

of the American people. He said: "As to the wealth
which the colonists have drawn from the sea, by their
fisheries, you had all that matter fully explained at
your bar. You surely thought these acquisitions of
value, for they seemed to excite your envy; and yet
the spirit by which that enterprising employment has
been exercised ought rather, in my opinion, to have
raised esteem and admiration. And pray, sir, what in
the world is equal to it? Pass by the other parts, and
look into the manner in which the New England people carry on the whale fishery. While we follow them
among the trembling mountains of ice, and behold
them penetrating into the deepest frozen regions of
Hudson's Bay and Davis' Straits, while we are looking
for them beneath the arctic circle, we hear that they
have pierced into the opposite region of polar cold;
that they are at the antipodes, and engaged under the
frozen serpent of the South. Falkland Island, which
seemed too remote and too romantic an object for the
grasp of national ambition, is but a stage and resting-
place for their victorious industry. Nor is the equinoctial heat more discouraging to them than the accu-
mulated winter of both poles. We learn that while
some of them draw the line or strike the harpoon on
the coast of Africa, others run the longitudes and pursue their gigantic game along the coast of Brazil. No
sea but what is vexed by their fisheries. No climate
that is not witness of their toils. Neither the perse-
verence of Holland, nor the activity of France, nor the
dexterous and firm sagacity of English enterprise, ever
carried this most perilous mode of hardy industry to
the extent to which it has been pursued by this recent
people; a people who are still in the gristle, and not
hardened into manhood."

This great interest was checked for a time during the revolutionary war, but was prosecuted with renewed ardor as soon as peace was declared. The waters of the southern seas have long been famous for the abundance of the black-headed sperm whale ; not away towards the frozen pole, but within the tropical circle and in the waters surrounding the islands of the South Sea. The fisheries in nearly every portion of these regions are followed with great success and profit, the black whale being taken as well as the sperm, in great numbers. These localities are favorite feeding and breeding grounds, the prolific animal life, the immense growth of the squid, the favorite food of the sperm, the immense number of marine animalculæ, the principal sustenance of the black whale, make a resort where mammals delight to make their homes.

Whale fishing has been so often and so well described, the minutest particulars being gone into, that the subject has become hackneyed, and is only alluded to here as one of the many prolific sources of industry and wealth offered to enterprise in the southern seas.

THE TURTLE AND ITS HABITS.

Many of the island groups of the South Sea are noted for the great number of turtle that frequent their shores. They are wonderful navigators, with very retentive memories, and, like the seal and sperm whale, do not make any new locations, but return year after year to the places of their birth, and make almost the same spots their feeding and breeding grounds. These exact habits have made the business of turtle fishing certain and profitable, not only as an article of food, but for the valuable oil they contain, and for the shell,

which modern processes have rendered of great value.
The tortoise, too, with its valuable covering, is much
sought after, and the shell worked into beautiful ornaments all over the world.

The food of the turtle is the sea-moss growing on
the coral reefs, and the young beche-de-mer, forming
their principal repast. The female, when about to
lay, which occurs once a year, chooses the time when
the moon is full, and is watched and guarded, during
this interesting process, by her mate, who lays "off
and on" just outside of the surf-line. Selecting a portion of the beach with a sunny exposure, she waddles
ungracefully to a point about ten yards beyond the line
of high tide, proceeds to excavate a place in the warm
sand something larger than her own body in diameter,
and in the center about two feet deep. Having arranged her nest in a satisfactory manner, she deposits
the eggs, about one hundred in number, and in size a
little smaller than a billiard ball. The nest is then
filled in with sand and levelled over, and great care
and attention exercised in obliterating all traces of the
sand having been disturbed. After taking all these
precautions, she hies herself to her mate, and they
swim contentedly away.

If she is disturbed during her preparation for
hatching, a retreat is made, and she will not be seen
again for nine days ; if again interfered with, she will
remain away for a like period ; and if still again disturbed, will seek some other favorite-spot or island,
and will not return to this particular breeding place
until the coming year.

If these places of incubation are watched, in about
a month the young turtle will be found digging their
way out of the nest and making for the sea. At this

time they are about the size of one of our silver dollars, and are quite lively and quick in their movements —which seems only a wise provision of nature, as they have many enemies to contend with.

THE TORTOISE.

The Galapagos are also celebrated for the great number of land tortoise that make their homes on those islands. They are of the genus *Testudo*, and are altogether inhabitants of the land; of little value, except as food.

The tortoise (*Testudo inbricata*), whose shell is so beautiful and valuable, is a species of sea turtle, and with similar habits. The shell of the tortoise covers the back in plates overlapping each other like tiling, and in its natural state is about one-eighth of an inch in thickness. It has the property of being molded in any form at a heat of 212 degrees, and retaining the given shape on cooling. Many tons of the shell are exported to Europe and America, where it is worked into the many ornamental and useful forms we meet with in the stores. This species is seldom found in the west longitudes of the Pacific.

SPONGE FISHERIES.

Sponges, classed by some writers as belonging to the marine species of vegetation, and by others to the marine animal kingdom, a species of the *zoophytes*, have long formed an important article of trade in all parts of the world. The Bahamas, in the West Indies, the Gulf of Mexico, the Mediterranean and Red Seas, the Levant, Green Turtle Bay, the Orient, with some

other localities, have for many years had almost a
monopoly of the sponge trade. I again refer to Mr.
Sterndale, whose personal experiences and writings
are of considerable interest. Among the profitable
industries of the coral seas, the collection of sponges
is not the least important. To fish for sponges with
success requires a certain degree of practice, as they
are very difficult to recognize in the water when in a
live state. They grow on the coral, and very much
in the crevices of it, and are not by any means conspicuous, as they look like a part of the stone.
When removed they are heavy, slimy, hard, and black
as tar. The best of them are of the form of a mush-
room, and are found from the size of a man's fist up
to two feet in diameter. In these latitudes they
usually lie within the lagoons, in water of a depth from
one to ten fathoms. They are inhabited by animalculæ, which in the process of cleaning are decomposed and washed away. In order to effect this object upon a sandy beach where the tide ebbs and flows,
a number of forked sticks are driven into the sand,
and upon them are fastened slender poles, as a sort
of frame-work; from these, sponges are suspended
by strings in such a manner that when the tide is in,
the sponges are floating in it; when the tide is out,
they are exposed to the wind and sun. In the latter
case, the animalculæ die and decay, and by alternate
sorchings and washings, the sponge becomes cleaned
and bleached, as well as softened, in consequence of
the removal of the glutinous creatures which had inhabited it. When prepared in this manner, the usual
rate of barter among the islands where they are
chiefly obtained, is four large sponges for one yard
of calico. I have found that they were greatly im-

proved both in color and softness by being washed in hot fresh water, which had been previously strongly impregnated with the alkali of wood ashes.

The better way has been found, as practiced on the Mediterranean and at the Bahamas, to use a weak solution of muriatic acid, which not only effectually frees it of animalculæ, but removes the last traces of lime adhering to the sponge.

.

CHAPTER XI'I.

ISLAND PRODUCTS AND RESOURCES.

Rocks are rough, but smiling there
Th' acacia waves her yellow hair,
Lonely and sweet, nor loved the less
For flow'ring in a wilderness.

                    MOORE, (*Lalla Rookh.*)

THE ROBBER CRAB.

I WAS a good deal interested during our voyage, in the many tales, legends and experiences so ready to the sailor tongue, some of which must be listened to and taken with a grain of salt. Yet at times I was able to verify what at first seemed to be some very hard tales. Thus, at Vanikoro, one of the Santa Cruz group, where we remained for nearly ten days, the great land-crab of the South Sea was met with, known here by the name of "Koviu." It was ascertained to be the Birgus latro, the Anamoura of the Crustacæ family, or, in plainer terms, and universally used in the Pacific, the Land or Robber Crab. Some of the species met with were over two feet long and about eighteen inches across. They live altogether on the land, seldom taking to the water, although perfectly at home in that element. Their nests are made among the roots of the cocoanut tree, and in the little caves and openings among the rocks and coral, and are nicely arranged for ease and comfort, being lined with the fibrous covering of the cocoanut. During the day they are seldom seen, selecting night for their peregrinations.

AS A GOURMAND.

Shrewd and cunning to a high degree, they seldom miss the hatching out of the young turtle, whose nests they watch with almost maternal solicitude. But for a somewhat different purpose—that of making a repast of the tender young turtles, as they are scudding for the water, and which they devour with the greatest gusto. I am told that one of the reasons of the extreme caution of the female turtle, when selecting places to deposit her eggs, is an instinctive fear of this highwayman. True, the crab does not care for the eggs, but, as the sailors say, when the young turtle are coming out, the "pirate never misses a trick."

AS A LOVER OF COCOANUTS.

Of course the "robber" does not depend upon this mode of getting a living at all seasons. Such opportunities occur only during the hatching season of the turtle, which is but once a year. Another of the favorite methods the crab resorts to for obtaining food, is the continuous growth of the cocoanut. Climbing the trees with great skill and a surprising quickness, he shears off the fruit from the stem, selecting such nuts as are nearly ripe. After obtaining about one dozen in this manner, and which are allowed to fall to the ground, he descends the tree, and, with his great

strong claws, strips the covering from the fruit, and
selects the end where there are several eyes or
openings in the shell, provided by nature for the easy
rooting or sprouting of the young tree ; then, forcing
some of the fingers of its great claws through these
into the nut, he deliberately hammers it on the rock
or coral until the shell bursts open, when the expected
feast becomes an easy matter.   Two or three generally serve for the morning's meal, the balance being
transported to the nest as a reserve.   When breaking
the shells they must exert great force and power, as
the reverberation of the blows, along the shore, may be
heard for a half mile.   All that I have related is performed with a method, foresight and skill; almost human.

A late writer says (now speaking of a larger
marine crab) : "I had heard of these giants, but I had
no idea that they attained this enormous size.   Though
this crab is the largest, it is not as powerful as the
famous palm-tree crab, of the islands south of Japan,
and in the Indian Archipelago.   The crab is called the
Birgus, and is a relative of the hermit crab, only it has
no shell, the plates on the abdomen being extremely
hard, and effectively taking the place of the shell that
is worn by others of the kind.   The Birgus is not a
water crab, living entirely upon the land, and going
down to the sea once a day, it is said, for the purpose
of moistening its gills.   They are generally found in
the near proximity of palm trees, upon the fruit of
which they live, and their burrows are generally placed
at the foot of the trees.   To give you an idea of the
number of cocoanuts the creatures eat, the Malays
come about twice a year and dig up their holes to get
the cocoanut husks that the crabs took in to make

their nests. Hundreds of pounds are thus obtained and made into mats, beds, and many other articles of household use.

STRENGTH AND TENACITY.

The most remarkable feature about these crabs was their enormous strength. One was placed in an ordinary tin cracker-box, where there was no opportunity of taking hold; but the next morning the box was found completely punctured with holes, actually bitten through by the sharp, biting claws of the crab; and in another confined in the same way, the top of the box was fairly twisted off. Having so much muscular power, natives naturally approach them with some caution, when attempting their capture. I was informed that on one occasion a party went out to a place somewhat famous for them, and arriving at night, with the expectation of trying for the crabs the next day. But during the night the party was awakened by the most terrific screams, and, rushing into the wood near at hand with rush lights, they found one of the natives swinging partly from the huge leaf of a cocoanut tree, and screaming as if he was being hung. For some moments they could not make out what the trouble was, but finally was sure the man was in the grasp of an enormous Birgus. The native had attempted to climb a palm tree, but had been seized almost immediately by a crab which happened to be clinging to the branch. Naturally the crab held on, and had almost pulled the hair out of the man's head before he was rescued.

The intelligence shown by these crabs is remarkable. They climb the palms, bite off a nut and allow it to drop, and thus break it open; and if they find a

nut on the ground, they have been known to take it to
the top of a tree and hurl it to the ground. Others,
and generally the large ones, have been observed to
beat it against a rock, and so break the shell. They
invariably commence to tear away the husk at the end
upon which is situated the two holes. When this is
done, with the great claw, they hammer the holes until
an opening is made, and then the body is twisted
around, and one of the small hind legs that will just
fit is introduced, the meat taken out bit by bit, and
then the shell is broken.   •

The crab is certainly a lowly creature, but it is
remarkably intelligent in some ways, and also cunning.
If you have ever tried to catch a wild lobster, you are
aware how many wiles they have to effect their escape
or delude their pursuers.

Some years ago the question was raised in London, whether crabs remained in the same locality year
after year, and finally it was resolved to test the question. So about a thousand crabs were caught and
marked in various ways, and taken a distance of twenty
miles, and put overboard, and in less than a week
hundreds of these marked crabs were caught on their
own grounds.

PLANTAIN OR BANANA.

Of this fruit Humboldt says: I doubt whether
there be any other plant that produces so great a
quantity of nutritive substance in so small a space.
Eight or nine months after the sucker is planted, it
begins to develop its cluster. The fruit may be gathered in the tenth or eleventh month. When the stock
is cut there is always found, among the numerous
shoots that have taken root, a sprout, being two-thirds

TEA PLANT OF CHINA—IN FULL BLOOM.

the height of its parent plant, and bearing fruit three months later. Thus a plantation of bananas perpetuates itself without requiring any care on the part of man, further than to cut the stalks when the fruit has ripened, and to stir the earth gently once or twice a year about the roots. A piece of ground of one hundred square metres of surface will contain from thirty to forty plants. During the course of the year, the same piece of ground (reckoning the weight of the cluster at from fifteen to twenty kilogrammes only) will yield 2,000 kilogrammes, or more than 4,000 pounds, of nutritive substance. What a difference between this product and that of the cereal grasses in most parts of Europe! The same extent of land planted with wheat would not produce above thirty pounds, and not more than ninety pounds of potatoes. Hence the product of the banana is to that of wheat as 133 to 1, and to that of potatoes as 44 to 1. The banana forms the principal food of these, as well as many other tropical countries, and the apathy and indolence of the natives in the *tierras calientes*, or hot regions, has been ascribed—and probably with good reason—to the facility with which it supplies them with a means of subsistence.

Again, the fruit is dried and pressed, after which it can be kept for a long time, forming a food not inferior to the dried figs of commerce.

BECHE-DE-MER FISHING.

In regard to the traffic in Beche-de-mer, for which there is such a demand in China and Japan, I have thought best to quote from Mr. H. B. Sterndale, who some years ago wrote many interesting papers on the islands of the Pacific:

Beche-de-mer, called by the Chinese *Tripang*, and by the Polynesian, *Rodi*, and in the South Sea and Caroline group, *Menika*, is of that species of mollusc classed as the *Holothurides*. It has the appearance of a great slug or leech, and like most other marine animals of the same type, lives by process of suction upon microscopic animalculæs. It has the form of an elongated sac of a gristly consistence, traversed internally by strong muscles. It grows usually to about eighteen inches long and somewhat less in circumference. The labor of collecting and drying the fish is performed partly by the crews of the vessels engaged in this business, who are commonly Polynesian natives, with the exception of the mate or trading master or interpreter, and such islanders as they bring along with them, if it be a desert or uninhabited place, or otherwise the aborigines whom they find in possession. There is one advantage in beche-de-mer fishing that upon the great desert reefs where it most abounds the fishers never need be idle. In calm weather they gather the red kind off the top of the reef just inside the foam of the breakers. In stormy times they dive for the black kind inside the lagoons.

METHOD OF FISHING.

From its size and color, it is plainly visible to a depth of at least ten fathoms, even when the water is much ruffled by the winds—the more so as it lives on the smooth white sandy bottom. The material required for the prosecution of this business is of the most limited character, merely a boat, a few axes to cut building material and fire-wood, a supply of long knives for all hands, and in some cases two or three try-pots, such as are used on board of whale-ships, with buckets and sluice-forks. The first preliminary operation is to build two houses—one for the curing of the fish, which is done by smoking, as bacon is cured in our own country; the other, for the purpose of storing it after being sufficiently cured.

When in proper condition it brings readily in China or Japan five to six hundred dollars per ton, with hardly a limit to supply or demand.

CONTRACTS WITH THE NATIVES.

The terms upon which the laborers are engaged for beche-de-mer fishing depends upon the circumstances of the case. "Beach-combers," who have native wives and families, commonly make up a party of their wives' relations and near neighbors, and remunerate them for their work by sharing a part of the proceeds. Adventurers who sail small vessels, and have no settled home on the islands to which the laborers belong, hire them for a specified time at a fixed rate of wages, under a written agreement, which is witnessed by their chief or king. Although in the majority of cases no one understands the document but the white men concerned in its concoction, yet the most ignorant of these natives are pleased to see a promise written down, there being to their untutored minds something sacred and binding connected with the operation. Here follows a verbatim translation of a memorandum of this kind between one "Uroroa" (that is Longbeard, a white man known to the natives by that name, as Polynesians generally invent a name from some physical peculiarity for any European whom they have dealings with) and certain people of Nukinivano:

"We, men and women of Nukinivano, whose
marks are put at the bottom of this paper, agree to
go with the captain Longbeard to the island of Gan-
net Cay, and to fish for beche-de-mer for six moons,
and to be paid each man or woman fourteen fathoms
of calico, or twenty-one plugs of tobacco per moon,
or other things as we like, such as knives or needles,
at a value as we have before agreed; and at the end
of six moons, to be returned to our home, if the wind
should be fair for us to come back at that time.
The chief, whose name is Dogfish, shall superintend
the work.   The captain Longbeard, shall tell the
chief Dogfish, what the people are to do, and Dogfish shall tell the people.   The captain Longbeard,
shall not beat any of the people.   The people shall
not fight among themselves, but if there be any quarrel among them, they shall refer it to the captain
Longbeard and the chief Dogfish.   If any one of the
people die, that which is due him or her shall be en-
trusted to the chief Dogfish, to be given to his or her
family.   The captain Longbeard shall supply to all
the people, for nothing, lines and fish-hooks, that they
may catch themselves food.   All food and fresh water
shall be taken charge of and fairly divided by the chief
Dogfish.   Twenty-eight days shall count for each
moon; out of each moon, shall be four days' rest,
that is to say, the people shall work six days, and on
the seventh day they shall do no work.   They shall
not lie to the chief Dogfish, or be lazy, sulky or dissatisfied.   There is no more to say."

Here follow the names of the people, with their
marks.   The contract they will keep to the letter, not
only performing the duties imposed upon them, but
adding deeds of bravery, kindness, and an obedience to

the orders of their employers, that might be copied
with great benefit in more civilized lands.

COCOANUT.

One of the great sustaining products of nearly all the
groups of the Pacific is the fruit of the cocoanut tree (*cocus
nucifera*), a species of the palm. The cocoanut is so
well known that but a passing allusion seems all that
is necessary. Yet its manifold uses, with that of the
tree on which it grows, if described at any length,
would fill a volume. In these latitudes it has a very
luxuriant growth, and gives to the lazy natives a never
ending supply, and at all times of the year something
to eat, drink and wear, with abundant material for
clothing and shelter. It may be found growing in the
valleys, on the hill and mountain sides and tops, and
on reefs and sandy shores, with its roots laved by the
waters of the sea. It grows to a height of sixty or one
hundred feet, from one foot to two feet in diameter,
bearing fruit seven years after it is planted, and lives
about eighty years. Each tree furnishes a hundred or
more nuts a year, while a wise provision of nature so
arranges it that the natives may find the nut in all of
its many stages of progress before ripening, and all on
the same tree. The nut, when fully ripe and ready to
fall, is covered with a thick fibre, that prevents it from
breaking or bursting when it strikes the ground.
From the upper end grows a flag or tuft that guides it
in its descent, and causes it to rest with its proper end
down, ready to take root and reproduce its species.
Again, this fibrous covering is impervious to water,
and should the nut fall in or be carried by the waves
or surf out to sea, it drifts and floats with the currents,

winds or tides, until cast upon some distant island, reef
or beach, to take root and grow, very often furnishing
subsistence and shelter to unfortunate castaways upon
otherwise barren islands.

An immense trade has been carried on for years
with China, Japan and Europe, in the preparation,
shipment and manufacture of cocoanut oil. For this
purpose the nuts are gathered, the covering taken off,
when they are piled in great heaps on rude platforms
about a foot above the ground. This is to prevent the
absorption of moisture from the earth, and consequent
germination. The nuts are allowed to remain in this
condition for several months, with frequent turnings
and handling. After drying sufficiently (ascertained
by average samples taken from the heaps), they are
broken open, and the "copra," or dried cocoanut kernel, is ready for shipment. In Europe it is consumed
in great quantities, the copra being pressed by machinery much like that used in extracting oil from flaxseed, the residuum being in the form of flaxseed cake,
and sold all through Europe as a valuable food for cattle.
For this latter reason, it has been found more profitable to ship in the form described, in preference to extracting the oil at the islands.

It is not an over-estimate to suppose that in a cocoanut plantation the trees will number sixty-four to
the acre, within a fraction of twenty feet apart, and
that each tree will produce one hundred nuts per annum. These will produce copra equal to five
hundred pounds per thousand, and from this, again,
twenty-five gallons of cocoanut oil can be pressed,

worth about sixty cents per gallon. At this rate, a cocoanut plantation would produce, of oil alone, very near $100 per acre per annum.

Some idea may be formed of the varied uses to which the cocoanut tree and fruit are put, when it is known that as many as thirty articles manufactured from them may be found in one ordinary English home. Where fabrics are not altogether made from the fibre, it yet enters in with other material. The oil is used in many ways, forming one of the principal ingredients in fine soaps and other similar manufactures. The fruit, while by itself considered by many a great delicacy, in combination forms an important ingredient in our pastries and candies.

The tree, when tapped, furnishes a pleasant, healthful drink, known as cocoanut toddy. Modern processes, though, have made this fluid into a rum, called arrack, and said to be very satisfactory to old drinkers in the way of strength and brain-entangling qualities.

CORAL (CORALLUM).

Up to 1751, the theory that coral was a vegetable growth (disputed by Feranto Imperato, the Neapolitan naturalist, as early as 1599) had been generally accepted. Even its scientific name, as applied to-day (zoophyte), given by Linnæus, indicates the struggle that sometimes takes place to throw light even into scientific minds. The name would place it in both the animal and vegetable kingdom, forming a rather curious combination for the industrious little insect to work under. In truth, coral is the stony frame which belongs to these animals, as a skeleton belongs to an individual of the higher orders of the animal kingdom.

The coral which has obtained world-wide celebrity, is that used as jewelry, known as *corallum rubrum*, found in the Mediterranean, the Barbary coast, the coast of Italy, and in some parts of Europe and America.

In general, the coral of the Pacific cannot be considered as valuable for jewelry, the order being of the coarser kind—curious and beautiful in its varied colorings and forms, but of no great intrinsic value—if we except a kind found along the shores of the island of Sumatra, and as we approach the Indian Ocean.

In the olden time, the manner of fishing for coral was nearly the same everywhere. That which is most commonly practiced in the Mediterranean Sea is as follows: Seven or eight men go in a boat, commanded by the proprietor; the caster throws his net (if we may so call the machine which he uses to tear up the coral from the bottom of the sea), and the rest work the boat and help draw in the net. This is composed of two beams of wood tied crosswise, with leads fixed to them to sink them; to these beams is fastened a quantity of hemp, twisted loosely round and intermingled with some loose netting. In this condition the machine is let down into the sea, and when the coral is pretty strongly entwined in the hemp and nets, they draw it up with a rope, which they unwind according to the depth, and which it sometimes requires half-a-dozen boats to draw. If this rope happens to break, the fishermen run the hazard of being lost. Before the fishers go to sea, they agree for the price of the coral, and the produce of the fishery is divided, at the end of the season, into thirteen parts, of which the proprietor has four, the caster two, and the other six men one each; the thirteenth belongs to the company for the payment of boat hire, etc.

CHAPTER XIV.

ISLAND PRODUCTS AND RESOURCES.

Cedar, and pine, and fir, and branching palm,
A sylvan scene, and as the ranks ascend
Shade above shade, a woody theatre
Of stateliest view.

MILTON, (*Paradise Lost*).

PAPER (PAPYRUS).

IT has often been a subject of wonder with those learned and ingenious persons who have written concerning the arts of the ancient world, that the Greeks and Romans, although they possessed a prodigious number of books, and approached very near to printing in the stamping words and letters and similar devices, should not have fallen upon the art; the first rude attempts at typography being sufficiently obvious, though much time and contrivance have been required to bring the process to the perfection in which it now prevails.

We owe the introduction of paper into Europe to the Arabians or Moors. There is some uncertainty as to the precise era of its first appearance, and we are unable to trace the origin of the precious invention, or even to imagine by what steps men were led to it. We cannot conceive how anyone could be tempted to pound wet rags in a mortar, to stir the paste into a large body of water, to receive the deposit on a sieve, and to press and dry it. The labor of beating rags into a pulp by hand would be as hopeless as it would be tedious and severe. It is true that paper was originally made of cotton, a substance less obstinate than linen and other rags, which are now commonly used. At present the fresh rags are torn into pieces by a powerful mill; formerly it was the practice to suffer them to rot, to place them in large heaps in a warm and damp situation, and to allow them to heat and ferment, and to remain undisturbed until mushrooms began to grow on them—so that, being partially decayed, it might be less difficult to triturate them. Nevertheless, the invention of paper is a mystery. The Chinese possessed the art of making paper and of printing, but we know not how long they have had them, nor whether the Mohammedans learned the former from them. The illiterate inhabitants of some of the islands of the South Seas were able to compose a species of paper, which they used in fine weather for raiment, of the bark of trees. The basis of paper being the vegetable fibre, it has been made of various substances, as straw, as well as rags.

(Notes from an old History of Paper-making.)

To describe the methods now in use for the manufacture of paper, with an account of the perfect machinery, taking place of human hands, in the various manipulations to turn out the beautiful paper now met with in nearly all parts of the world, would take up a volume. On the other hand, with all our perfect manufacturing appliances, we lack the natural vegetable growths of just those piths, pulps and barks, that nature so abundantly scatters broadcast throughout the islands of the Pacific. Paper exhibited at the last Exposition in Paris, manufactured in Japan, it is said from the bark of the mulberry, being in truth the *Broussonetia*, the Paper Mulberry of Japan, the East Indies and the South Sea Islands, excited general admiration. Paper from that country that I have inspected very lately in San Francisco, is far superior in texture, beauty and durability, to any of the brands made from English linen. Samples from the Phillippine Islands, made from the *abaca*, and others of the *musa* (banana) plants, show fully as fine and strong a texture, but lacking the satiny gloss of surface, like watered silk, seen in the samples from Japan. The vegetable growth furnishing the textile fabrics in all its many varieties, is to be found in wild abundance on nearly all of the Pacific islands. The gathering of the raw material, and its export to Europe and America for its more perfect manufacture into the manifold forms of paper, would naturally lead to a vast business in the textile fabrics alone, that would result in many profitable industries.

CINNAMON (CINNAMOMUM ZEYLANICUM).

Cinnamon is of the same species as the laurel. The tree is of small growth, and evergreen. In the island of Ceylon, where the finest qualities are produced, it is cultivated in a large way, and forms no inconsiderable portion of the princely revenues received from the products of that island. It is claimed by many authorities to be indigenous to the soil of Ceylon. In any event, the product is far superior to that of any other part of the world, although many of the island groups cultivate and produce it in abundance, but of much inferior quality. The gardens where the best cinnamon is grown are managed on the coppice system, the tree being cut down almost to the roots, and the young shoots, some six or eight only, allowed to grow. At the age of two years the shoots have reached a height of about six feet, with a diameter from one to two inches. These are cut, and the bark peeled off, being afterwards cleaned and scraped, when it is rolled and dried, tied in bundles, and is then ready for the market.

The ordinary cinnamon as we find it in the marts of America, is *cassia bark*, a species of the cinnamomum, grown extensively in China, Japan and all the islands of the Eastern Archipelago. It is sold in immense quantities, particularly after it has been ground, when it is hard to distinguish it from the true spice.

INDIGO.

Another of the valuable plants having indigenous growth in the island groups of the Pacific, particularly in Japan, the Phillippines and Java, is indigo. Derived from the maceration in water of the leaves and twigs of the plant, *Indigofera tinctoria*, and the *Indigofera Anil*, with its after precipitation from the liquid form into that met with in commerce, it may be justly termed one of the valuable island products. The indigo from the island of Java, the result of the rude methods of manufacture resorted to by the natives, is the finest in the world, the plant seeming to thrive best when of island growth. Its manifold uses as a drug, as well as in the arts, together with the rather complicated processes necessary for its extraction from the plant, would require at least a separate chapter. As a product of the island world, it is of considerable importance.

TEAK WOOD.

Among the many valuable trees, growing so profusely on the Islands of the Pacific, may be cited Teak, or Indian Oak, the product of the *Tectona grandis*, a large forest tree, growing in the dry and elevated districts in the south of India, the Burman empire, Ava, Siam, Sumatra, Java, Borneo, New Guinea, etc. Teak is by far the best timber in the East; it works easily, and though porous, is strong and durable. It is easily seasoned, and shrinks very little. It is of an oily nature, and therefore does not injure iron. Mr. Crawford says that in comparing teak and oak together, the useful qualities of the former will be found to preponderate. It is equally strong, and somewhat more buoyant. Its durability is more uniform and decided; and to insure that durability, it demands less care and preparation, for it may be put into use almost green from the forest, without danger of dry or wet rot. It is fit to endure all climates and all alternations of climate. The teak of Malabar, produced on the high table land of the south of India, is deemed the best of any. It is the closest in its fiber, and contains the largest quantity of oil, being at once the heaviest and most durable. This species of teak is used for the keel, timbers, and such parts of the ship as are under water; owing to its great weight, it is less suitable for the upper works, and is not at all fit for spars. The teak of Java ranks next to that of Malabar, and is especially suitable for planking. That of Sumatra, Borneo, New Guinea,

etc., are of equal value, and their great forests teem
with such an abundance as to be able to supply
the ship-building material for the navies of the world.
The Rangoon or Burman teak, and that of Siam is
not so close grained and durable as the others. It is,
however, more buoyant, and therefore, best suited
for masts and spars. Malabar teak is extensively
used in the building-yards of Bombay. Ships built
wholly of it are almost indestructible by ordinary
wear and tear, and instances are not rare of their
having lasted from eighty to a hundred years ; although
they are said to sail indifferently, but this is probably
owing to some defect in their construction, and not to
the weight of the timber. Calcutta ships are never
wholly built of teak ; the timbers and frame-work are
always of native wood, and the planking and deck
only of teak. With this timber, in combination with
the pine of Oregon and the redwood of California,
vessels could no doubt be constructed superior to
anything being built in our ship-yards at the present
time.

RICE.

One of the great food staples of India, China,
Japan, and the westerly islands of the Pacific, is rice.
It is among the most valuable of cereal grasses—the
oryza sativa of botanists. It forms the principal part
of the food of the most civilized and populous Eastern
nations, being more extensively consumed for that
purpose by the people of those countries, than any
other species of grain. It is too well known to require
more than a place here as a product. The quality of
the grain grown is not equal to that produced on the
low, marshy grounds in the Carolinas of America—it
having no equal.

SILK (SERICUM).

The art of rearing silk-worms, a species of cater-pillar or larvæ of the genus *phalæna*, and of unraveling the threads spun by them in forming their cocoons, dates away back in the dim pages of Chinese history. Its first introduction from China into Rome was about the time of Pompey and Julius Cæsar. The great distance of China from Rome, the journeys of the caravans overland through the Persian Empire, caused a high price to be placed on silk, bringing in the earlier periods its weight in gold. The art of rearing the worms gradually extended over the countries of Europe, being introduced in France under the reign of Louis XI in 1480, and into England at about the same period. The manufacture of silk was begun in Lyons in 1520, under Francis I. The art gradually extended itself over France, and in such esteem were its promoters held, that silk manufacturers who had pursued the trade for a period of twelve years were rewarded with a patent of nobilty by Henry IV.

Rearing the silk-worm, with the cultivation of the mulberry (*moraceæ*) tree in its many varieties—the leaves of which serve as food for the worm—has been reduced to a fine art in India, China, Japan, the Phillippines, and some of the islands of the Eastern Archipelago, and other islands of the South Sea, forming one of the most valuable productions of those places, and forms no inconsiderable portion of our commerce with the localities named.

PINEAPPLE (ANANASSA SATIVA).

This delicious fruit is native to most of the tropical islands of the South Sea, and like that grown in the hot-houses of England and America, its quality as a fruit is altogether dependent upon the care exercised in its cultivation. In its wild state, about the only condition in which it produces a reproductive seed, it is hardly ever sought after as a food, but rather for the long, fine fiber contained in the leaves. There are as many as fifty varieties, not all of them bearing a palatable fruit, even when cultivated. That thought the most of, in the Phillippine Islands—not as a fruit, but for its fiber-producing qualities—grows in the wild state, and is known to botanists as the *Bromelia pinguin*. This particular plant throws out leaves from three to sometimes eight feet long, which abound in fiber of great strength and durability in the older plants, while in the leaves of the younger growth a fiber is found that the natives work into all the delicate forms, gossamer and cobweb like, and in such delicate and beautiful designs as not only to always astonish the traveler, but to invariably bring, when in the form of veils, handkerchiefs, etc., many times their weight in gold.

MANILLA HEMP.

The textile fiber of the *abaca* palm, of the family of *musas*, to which the banana and plantain belong, is found native in a great many of the island groups of the South Sea, but probably is better known and grows in greater luxuriance in the Phillippines, where the manifold uses the fiber is put to, in the manufacture of the most delicate laces, veils, handkerchiefs, to the coarsest cables used by ships, has made the name of the hemp world-wide. The thousands of tons of the raw material shipped from the Phillippines every year, and to nearly every part of the world, bear evidence as to its value and the continually increasing demand for the fiber. The luxuriance of plant growth throughout the islands of the Pacific, may yet be taxed to supply the growing demands of the world, for products lavished by nature on these sunny lands. From this same and kindred plants, a great quantity of paper is made, and the fiber is spun and woven alike by the natives into a superior cloth for clothing, or into a heavier material for sails, mats, bagging, etc.

PEPPER (PIPER).

The fruit of the climbing shrub or vine (*piper nigrum*) is native and cultivated in many of the tropical countries. Although a spice, apparently used in small quantities, yet in the aggregate, thousands of tons of it are produced and exported from the Pacific islands each year.

Java, Borneo, Sumatra, the Phillippines and the Molluccas furnish the little pungent berry in abundance. Where not native in the grand old forests of the islands, or when not supported by trees, the plant is cultivated in a manner very similar to our hop fields. The black and white varieties are the product of the same plant, the latter simply being put through a bleaching process, in water or by chemicals, and results in the white pepper of commerce. Pepper is not at all a product of the South American pepper tree, much used in our country for shade and ornament; the berry produced being similar in appearance to that of the pepper plant, together with the name, the erroneous impression sometimes prevails that the pungent product is from this tree. The effect of the pepper tree berry on the system is somewhat different

14

from that of the true pepper. Red pepper, also a great island product, is from the plant—genus *solana-ceæ*, or nightshade family, and is grown in all parts of the world. It is native to tropical countries, and in the islands of the Pacific grows in the greatest luxuriance. After ripening on the plant, it is picked, dried and ground, furnishing the Cayenne pepper of commerce.

GUTTA PERCHA.

Gutta Percha is the name given by the Malays to the tree belonging to the natural order *sapotaceæ*, and to the newer genus *isonandra*, is found in the greatest abundance in the forests of Borneo, Sumatra, Java, and, in fact, throughout nearly all of the island groups where the forestry is abundant. The tree ranges from two to eight feet in diameter, and reaches a height of sixty to eighty feet. The timber is of great value, and is used by the Malays in many of their manufac-tures. The sap from the tree, after being reduced to the form of a gum, with its valuable property of be-coming plastic in hot water, so that it can be molded up into any form, retaining the shape when cooled, was known to the Malays probably for ages. This property, from which so many useful advantages have been derived, seems to have remained unknown to our people until about 1842 and '43, when specimens of the gum were forwarded to England, and some time transpired before it was brought into practical use. Gutta percha differs very materially from india rubber (also one of the bounteous products of the islands), in being elastic only in a very slight degree. The plants are very different. The india rubber, although growing a foot or so in diame-

PARADISEA PAPUANA.—Bird of Paradise—Young male, emerald throat.   From the Island of New Guinea.

ter, grows like a vine, and is often found twined around or clinging to the trees of the great island forests. Again, there is the important difference in the two gums, that rubber requires a chemical preparation with some of the earths, or to be mixed with certain proportions of metallic oxides, to make it harder after heating and molding, before it will retain the shape desired, becoming then *vulcanized* rubber.

SCREW PINE (PANDANUS).

This tree, much valued in the Pacific, is native to most of the islands, where it grows in the greatest abundance. It is among the first of the plants to appear on newly formed or forming islands, and with its spreading roots, often raised above the ground and supporting the main trunk on their stems, it acts as a dam and barrier to encroaching waves, and performs an important part in collecting and retaining the drift and debris, that assists so materially in the first plant growth of islands. Its leaves, growing generally from the ends of the main branches, spreading from the trunk, grow similar to those of the pineapple, whence its name; but unlike the latter, it is a tree growing from twelve to forty feet high. The many ways that the bark, timber and the strong fiber of its leaves can be used, makes it highly prized by the natives.

RESINOUS GUM TREES.

The great forests of Borneo, Sumatra, Java, Celebes, New Guinea, etc., teem with an almost endless variety of trees that furnish the liquid resins so valuable as a base for our varnishes, while the ground itself

supplies many forms of the oxidized fossil kinds,
such as copal, amber and others. From the Fiji
Islands, where the natives use a liquid resin as a
coating or glaze for their pottery, to the more advanced usages of the Japanese, in their beautiful
lacquer ware, also the results of resinous products, a
vast field in this line alone is spread out, offering
ample room for the employment of the capital, enterprise and skill of thousands of our unemployed
people.

GENERAL REMARKS.

In glancing with me in this general way at some
of the valuable island products, the intelligent reader
will no doubt agree with me in the assertion that it is
but a glance. That a volume could be written on
valuable products alone, and still another on their
manifold uses, and again another on the mechanical
appliances necessary for their more perfect manipula-
tion in manufactories.

CHAPTER XV.

OCEANIC ETHNOGRAPHY.

See him from nature rising slow to Art!
To copy instinct then was reason's part :
Thus then to man the voice of nature spake—
Go, from the creatures thy instructions take ;
Learn from the birds what food the thickets yield,
Learn from the beasts the physics of the field,
Thy arts of building from the bee receive ;
Learn of the mole to plough, the worm to weave,
Learn from the little Nautilus to sail,
Spread the thin oar, and catch the driving gale.

POPE (*Essay on Man*).

OCEANIC RACES.

WE shall use the term Oceanica in the sense in which it is applied by many writers on Ethnography, as describing all the land comprised between the coasts of Asia and America, including the East Indian Archipelago, the many smaller clusters of the Pacific, and the continent of New Holland.

The whole subject of the distinctions in race among the wild inhabitants who have settled on these countless islands—the "nomads of the sea," as Professor Muller calls them—is even more intricate and involved than the differences among the nomads of the land. The languages of many of the tribes have never even been compared, and some of them are scarcely known at all; so that all conclusions must necessarily, as yet, be very doubtful, and liable to much change hereafter.

There are at least two very different schools on this subject, each represented by high authority. One led by the celebrated William von Humboldt, assigns but two, or at most three, races of men to this immense range of inhabitable land—namely, the Malay, the Polynesian, and a race of Oriental negroes.

The other, represented by a scholar of great ability, Mr. J. Crawford, divides the inhabitants of Oceanica into five brown races, with lank hair, distinguished by varieties of language, and eight races of Oriental negroes. The tendency, however, of all late investigation, is toward the unity of these varieties, and modern conclusions approach those of Humboldt much more than those of Crawford.

Oceanica may be divided into five great divisions: *Malaisia*, or the East Indian Islands, together with the peninsula of Mallacca, inhabited by the Malay race. Of these islands, the most prominent are Sumatra, Java, Borneo, Celebes, Mollucca, Sooloo, and the Phillippines.

Melanasia are the islands inhabited by a dark race with woolly or frizzled hair, comprising New Guinea, Aroo, Mysol and others, together with New Britain, New Ireland, the Solomon Isles, and New Hebrides.

Australia, or New Holland, a vast island, sparsely peopled by a black race with straight, smooth hair.

Micronesia, a long range of little groups of islands and strips of coral rock in the North Pacific, east of the Phillippines, including the Pelew, Caroline, Ladrone, Bonabe, and numerous other islands, from 132 deg.

east longitude to 178 deg. west, and from 21 deg. north latitude to 5 deg. south.

Polynesia, or the islands in the East Pacific, occupied by a race kindred to the Malay, of which the best known are the Navigators, the Friendly, Society and Sandwich Islands, together with New Zealand.

The great natural peculiarities of this quarter of the globe, which have determined the divisions of race and family, have been its insular character, the periodicity of its winds, and the malarious climate of some of the islands; while the existence of a people on its western border, with a highly flexible and euphonious language, and gifted with much enterprise (the Malay race), has affected the ruling stock through all this wide region. These nomads of the sea, whenever desiring adventure or seeking commerce or plunder, or driven forth by defeat or hunger, had only to put themselves and wives, with their few utensils, into their light canoes, and trust themselves to the prevailing trade winds, and they were certain finally to land on some new island, where they could either intermingle with the old inhabitants or form a new community. It is thus that the almost countless islands, from the Phillippines to Easter Island, through eight thousand miles of ocean, were peopled by a similar race.

There were certain of the islands which only admitted of the habitation of the black tribes, owing to the highly malarious character of the climate, and upon them especially these tribes are found.

CLIMATE.

The climate has probably protected them against the assaults of the more organized nations. Whether

they were the original settlers is impossible to determine. Their usual position, on the mountains in the
interior of an island, would indicate an earlier habitation. Possibly, as some ethnologists have supposed,
their appearance here may date back to an immense
antiquity—before all the islands were separated one
from another or from the Asiatic continent;* while
their color and power of resisting malarious influences
may be due to the gradual accumulation and trans-
mission of advantageous changes, adapting them to
their circumstances through vastly extended periods
of time.

Judging from the gradual change in language and
customs, as well as from other indications, the great
movement of the Oceanican people must have been
from the west to the east—against the prevailing trade
wind; and no investigations show that even now, at
peculiar seasons of the year, there are regular winds
blowing from the west which drift the natives hundreds
and thousands of miles.

One great link has perhaps been discovered by
Professor Muller and others, showing the connection
between the nomads of the sea and the nomads of the
land, in their investigations into the Tai and Malay
languages.   *   *   *   These generic expo-
nents or numerical affixes are entirely peculiar to those
languages. Many other evidences are adduced of the
relation between the languages of the islands and the
Asiatic continent; so that, if this vast connection be
fairly established, the language of a vast portion of
Oceanica may be included in the great Turanian family.

*Both Dana and Hale notice evidence of a gradual subsidence of
the land, even in the historic period; the ruins of temples on Bonabé,
for instance, being found partly submerged by the sea.

THE MALAYS.

Besides the large islands, which have already been spoken of as occupied by this family, they hold also the small islands south of the Phillippines, up to the west coast of New Guinea, and those on the east point of Java and Sumatra, up to the Straits of Mallacca. Their language, which is found purest on the Phillippines, is one of the most widely extended of Asia, traces of it being discovered from Madagascar to Easter Island, and from Formosa to New Zealand, over 70 deg. of latitude and 200 deg. of longitude. This race has for ages possessed the knowledge of letters, worked metals and domesticated useful animals, and has led the commerce and enterprise of the Pacific Ocean. The flexibility of its tongue has made it everywhere the medium of communication, and even in Madagascar, at 3,000 miles distance, Malay words form one-fifty-seventh of the vocabulary of the islanders. The Malay conquest and settlements after the remote emigration from the continent, are supposed by Crawford to have begun from the center of Sumatra, and to have extended from the Malay peninsula and the coasts of Borneo. Their influence was only excluded from two quarters by different causes—from the Asiatic shores, by the superior Chinese civilization already prevailing there, and from Australia, by the great degradation of its inhabitants. Physical objects alone prevented their reaching the coasts of America. The Malay language shows that it has been acted upon by both Indian and Chinese influences.

The Malay bodily type is described by Prichard as Indo-Chinese. The nose is short, but not flat, the mouth large and lips thin, cheek bones high, and face broadest at that point, the complexion yellowish. The form is squat, and height only about five feet three or four inches.

THE POLYNESIANS.

The second great race, of similar physical structure and language with the Malays, and undoubtedly of the same origin, are the Polynesians. The islands especially occupied by this people are those lying between New Zealand and Easter Isle, north, up to the Sandwich Islands, and west, as far as the Fiji and New Hebrides. Mixtures of this with other races are found all over the islands of the Pacific. They were for centuries a half civilized people, and have possessed a well established government, together with religious doctrines and usages, and a sacred language unintelligible to the people, as well as a system of ecclesiastical authority. They exhibited skill in various arts, and were bold and experienced sailors. They had no writing, but possessed many legends and traditional poetry. Yet they and their kindred, the Malay race, have the infamy of being the principal and almost the only race indulging habitually in canibalism.

Physically, the Polynesians are placed among the class of light-brown complexion verging to white. They are described by Hale as above the middle height, well formed, with thick, strong, black hair, slightly curled, and scanty beard; the head short and broad, and higher than most races in their stage of development, with a remarkably flat posterior head, like that of the American Indians. In disposition they are represented as good-humored and fickle, and very ready to adopt new usages.

The Polynesian language, Hale supposes to spread especially from Bouru, the easternmost of the Malay islands.

The whole number of the Polynesians proper is less than 500,000.

From the evidence of language, Mr. Crawford concluded that there was, in the ante-historic times, a great Polynesian nation, whose speech lies at the basis of all the various Malay and Polynesian languages at the present day. This people—judging from the records preserved in the words they have transmitted—had made some progress in agriculture, and understood the use of gold and iron; were clothed with a fabric made of the fibrous bark of plants, which they wove in the loom, while knowing nothing of the manufacture of cotton, which they acquired afterward from India. They had tamed the cow and buffalo, and possessed and fed upon the hog, the domestic fowl and the duck.

The massive ruins and remains of pyramidal structures and terraced buildings on the Pacific Islands, are probably from this primeval race.

THE MICRONESIANS.

Micronesia, as was before stated, embraces a long range of small islands in the North Pacific, east of the Phillippines, including the Pelew, Ladrone, Bonabe and others, from 132 deg. east longitude to 178 deg. west, and from 21 deg. north latitude to 5 deg. south.

Owing to the peculiar position of these islands,

they are exposed to winds blowing from various quarters, so that the emigration which settled them would naturally be from many different sources. In physical type, the people are of reddish brown complexion, rough skin, and high, bold features; the head is high compared with its breadth, hair black and curled. They show skill in various arts, and, in Hale's view, give indications of having descended from a higher to a lower civilization. In advance of the Polynesians, they possess the art of varnishing and weaving; they also understand steering by the stars. The practice of tattooing is observed, not only for decency or ornament, as with other tribes, but for the purpose of distinguishing clans and memorizing events. Their government is more intricate than that of the Polynesians, and their religion is different, resembling more that of Eastern Asia, and recognizing the worship of parents. *Taboo* is not in use. On some of the islands, as Bonabe · and others, architectural ruins of a remarkable appearance are found. The language of Tarawa contains a mixture of Polynesian and Melanesian, or Papuan, but on the whole, it is uncertain if there is a distinct Micronesian race.

THE MELANESIANS.

The black tribes of Oceanica present a difficult subject to the student of races. Not enough is known of their languages, to affirm either as to· their origin or their division.

They are found first in the west, on the Andaman Islands, between 10 deg. and 14 deg. north latitude. These Melanesians, or Negrillos, are considered by Prof. Owen as the lowest of mankind. They have no tradition or history; no inventions, except door-mats, and bows and arrows; no agriculture, and their habitations are the rudest and most primitive. Both sexes go naked without shame, and families and wives are in common. According to the same authority, the Andamans have no notion of Deity, or spiritual beings, or a future state; an assertion which does not seem easily proved. They are not cannibals, but show a great hostility to strangers. Neither skull nor teeth present the characteristics of the lowest African tribes. Prognathism is no more common than in the most of the South Asiatic peoples. The hair resembles that of the Papuans and Australians, as well as of the lower African negroes. They approach the orangs and chimpanzees in their diminutive stature, but show the well balanced human proportion of trunk and limbs. Latham states that there is a very evident link of connection between the language of the Andamans and the monosyllabic Burmese.

The black tribes next appear in the Nicobar Islands, then upon the mountains of Mallacca, where they are called Semangs, and in the Phillipines, where, under the name of Negritos,* they number about 25,-000.

On Luzon there are 3,000 of them under the Spanish rule. On Ceram a tribe of them is found so

---

*The Negritos are said by Bowring to possess a remarkable facility in the use of their *toes*, and their feet are marked by a greater separation of the toes than usual. They can descend the rigging of a ship head downward, clinging with their feet. They are slight in form, agile, small and thin, with handsome face, and dark copper complexion. The hair is black and curly, head small and round, forehead narrow, eyes large and penetrating, and veiled by very long eyelids, the nose of medium size, slightly depressed, mouth and lips medium, teeth long.— (*Sir J. Bowring's Visit to Phil. Islands.*)

low as to live in trees instead of huts. A wild race
of blacks is supposed also to occupy the interior of
Borneo, though there is not full evidence of it.

GENERAL CHARACTERISTICS.

Crawford supposes that there is but one race of
Oriental negroes, as these blacks are called, north of
the equator, and two races south in the Malay Archipelago, and in New Guinea. Of these latter, one has
the negro features, but not in the extreme. The hair
is frizzled, long and bushy, skin of lighter color, forehead higher, and the posterior head not "cut off," as
it were. The nose projects, the upper lip is larger,
and prominent, and the lower very projecting. The
other race he distinguishes by its lank hair.

The more general conclusion now is, that there is
but one race of Oriental negroes, even including the
black Australians, and the inhabitants of Van Dieman's Land. Latham doubts even the existence of
the negro tribes in the smaller islands of Melanesia.

The Australian languages are more like the Malay
and Polynesian than they are like anything else. There
are often, he allows, greater approaches of the black to
the brown tribes in language, than the received physical
divisions would justify.

The black tribes are not considered by travelers
as inferior in capacity to the brown, but they are peculiarly wild and impatient of control, and thus not
easily organized, so that they readily fall under the
power of the Malays. It is not found to be true that
they disappear before the advance of civilization in the
Eastern Ocean. On the contrary, in some islands,
even the most civilized, they have increased; but the

great cause of their decrease is to be found in the bitter hostility and superior organization of the Malays and Polynesians.

Without the knowledge of their languages, these physical divisions are not sufficient to determine the origin or the divisions of the race. The probability is that these black tribes are offshoots from the ancient black races of India and Asia, scattered widely, by the conquest of others, or their own pursuit of plunder, over the Pacific islands. . A black tribe is known to exist on the mountains between Cochin China and Cambodia, called the Moys, which may be a portion of their ancestral people. On some of the islands which the black nations settled, they were extirpated, or were driven to the mountains, where they are still found; on others the malarious climate defended them from foreign encroachment, and on others they became mingled with a different race. Many of the Melanesian tribes present great mixtures of blood.

The Papuans, who are distinguished by spirally twisted hair, frizzled and dressed by them in a huge mass above the head, are a cross of the dark races with the Malays. The eastern islands, as Tanna and others, show Polynesian blood. Timor contains within its limits every variety of color and hair. The Fijis* are probably a mixture of Papuans and Polynesians. In their mould, they are said by Mr. Williams to be decidedly European, with very large and powerful frames. The face is oval, profile vertical, nose well shaped, but the hair frizzled and bushy. The com-

---

*The Fiji Islands, Mr. Williams supposes to be the point where the Asiatic and African elements, among the Polynesians, unite.

H. C. von der Gabelentz finds evidence of the mixture of Polynesian and Melanesian in the Fiji language.

plexion is between the black and brown—sometimes
almost purple. The nearest approach to the negro is
on the island of Kandavu. The Fijis resemble the
blacks in their use of the bow and the manufacture of
their pottery, and the Polynesians, in the making of
their paper cloth, the preparation of *kava*, and the
practice of tattooing. The language contains one-fifth
of Polynesian words, and four-fifths unlike any other
tongue. The aborigines of Van Dieman's Land are
classed by some among the Papuans. The Melanesians are notoriously sullen in disposition and deficient
in enterprise, and manifest a different temperament
from either that of the Polynesians or Africans.

The prominent distinction between the languages
of the negro and brown races, Crawford states to be
that the first contain more consonants in proportion to
vowels, and more harsh combinations of consonants,
than the latter.

Gabelentz has made a careful investigation of the
dialects of many of the Melanesian tribes. Those, for
instance, of the inhabitants of the Fiji Islands, of Anna-
tom, Eromengo, Tanna, Mallikolo, Mare, Lifu, Bala-
dea, Bauro and Guadalcanar.

His deliberate and carefully formed conclusion is,
that all the Melanesian languages, though disintegra-
ted and apparently separated from one another, owing
to the barbarism and isolation of each of the tribes, do
yet belong to one stock. He is also of the opinion
that, both in roots and in many grammatical peculiarities, there are numerous remarkable resemblances
between the Polynesian and Melanesian; so that the
hypothesis of their common origin is a highly probable
one.

If this be hereafter more fully demonstrated, the

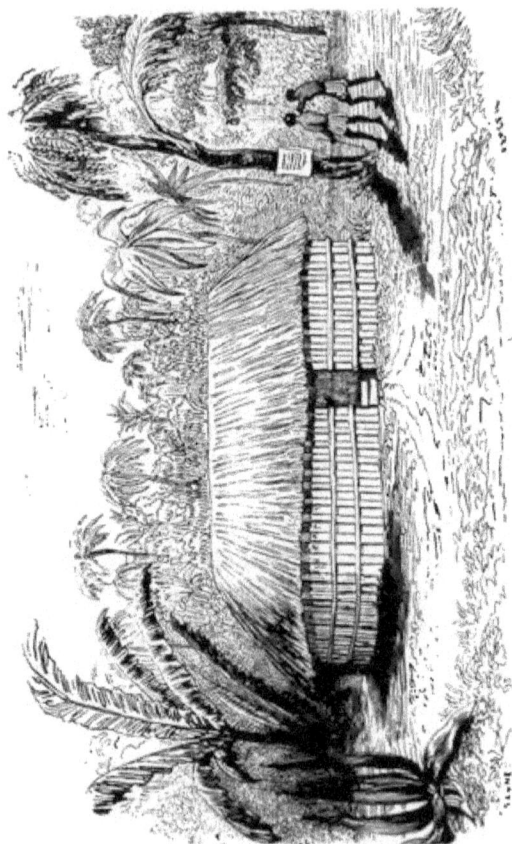

A NATIVE HUT AND GROUNDS UNDER "TABOO,"—CAROLINE GROUP.

whole vast population of brown and black peoples—
the Malays, Polynesians and Melanesians—may be
referred to one source, and in all probability be joined
with the Turanian races of Asia.

THE AUSTRALIANS.

The inhabitants of Australia and Van Dieman's
Land, belonging to the black races, are pronounced to ·
be almost the lowest of mankind. They have no government, and their religion consists only of the most
childish or debased superstitions. Their physical type
seems a cross of the Malay and the African, the most
distinguishing feature being the long, fine, wavy hair,
like the hair of a European. The evidence with
reference to their physique is quite conflicting. Many
of them are said to show a deficiency of bone in their
structure, and some tribes are represented as so
degenerated physically as to resemble *cretins*, and to
be in process of extinction. On the other hand, Pick-
ering states that one of the finest types of muscular
frame, and the most classic mould of head he has ever
beheld, he saw among the Australian natives. He
speaks of them as active, strongly formed and stately.
Various physical types probably exist among them.
In general, the features are as follows: The forehead
is narrow; mouth large, with thick lips; the nose,
depressed and widened at the base, but often aquiline;
the beard thick, the form slight, though well propor-
tioned, and color black. The number of these blacks
in Australia is said to be about 200,000. They are
supposed to be all of the same stock, though this conclusion is derived more from a resemblance discovered
in a few words, than a close comparison of grammar.

Not a Malay word is found in their language. Of their character, a competent witness (Rev. Wm. Ridley) says, that they are deficient in forethought and concentrativeness, but that in mental *acumen*, and in quickness of sight and hearing, they are superior to the whites. They are generous, honest to one another, and often attentive to the weak and the aged, though cruel to women. Notwithstanding their barbarous condition, there exists among them a very strict division of castes, and a certain kind of priesthood.

INTELLECTUAL CAPACITY.

It is interesting to know what capacities the lowest tribe or race of the human family may show. We learn, from quotations of a recent report to the English Government on this subject, that the Australian negroes show minds quick and keen—"rather like a treasure sealed up, than a vacuum." Their perceptive faculties are remarkable—far superior to those of Europeans—while, as might be expected, they are deficient in the reflective powers. As a consequence, the children are found to learn an external study, as geography, with great readiness, though showing much inaptitude for an abstract study like arithmetic. Mr. Parker, a visiting magistrate of the school in Mt. Franklin, says, that the native children manifest just as great capacities for improvement as do English children, and that the main obstacle to their elevation is from moral rather than physical causes.

The numerals of the Australian languages rarely reach five, and generally stop at three. Some affinities have been discovered between them and the Tamul.

We have classed the Tasmanian tribes (of Van Diemen's Land) with them, but the basis for classification is as yet extremely uncertain.

The great difficulty in determining the races of Oceanica is, that the tendency of a nomadic people to continually form new words and new languages, as they found new colonies, is here intensified by the separation which the sea naturally causes. There is something, too, in the disposition of the black races which has doubtless increased this tendency to disintegration. Crawford, who may have exaggerated in this particular, states that there are forty languages on the little island of Timor, and many hundreds in Borneo.

Nearly all writers allow that climate and circumstances have produced the most marked effects here on persons of the same race. Among the Tahitians and Maorians, for instance, the lowest castes are found nearly as black as negroes, and with crisp, woolly hair, while the higher (the chiefs and others), less exposed to the sun and the influences of the weather, resemble Europeans both in features and complexion; though both, there is every reason to believe, belong to the Polynesian race. Similar differences are observed on New Zealand among the blacks.

The Semangs, the blacks of Mallacca, are brown where not exposed to the sun, and in language and character have so strong a resemblance to the Malays, as to be considered by many, a tribe of that race.

The points of resemblance between the Polynesians and the Central American Indians are so striking, as to induce many writers to assign the same origin to both peoples.

The Asiatic origin of the Malay-Polynesian races

seems to us clearly indicated, so that all these resemblances cannot be considered in this connection.

(Brace: The Races of the Old World.)

INFLUENCE OF OCEAN CURRENTS.

I have quoted thus freely from the works of Mr.
Brace, with the object, not only of proving the origin
of the island races, but with the view of tracing the
source (only in a general way, however) of a portion
of the inhabitants of North and South America, as well
as the islands of Oceanica.   In another portion of this
work, the ocean currents of the Pacific have been alluded to as the great highways over which the Asiatics
voyaged, to people the New World.   Mr. Brooks, in
his work on Japanese Wrecks, accompanied by a map
of the Northern Pacific, showing the location of wrecks
discovered within a few hundred years, clearly shows
the influence of the northern current.   They are traceable from a short distance from Japan to Kamptchatka, the Aleutian Isles, Alaska, British America,
Oregon, California, Mexico, the Equator, and westerly
into the islands of the South Sea; always being found
in the line of the Japanese Black Stream.   It is
doubtful if any of these wrecks were found following
the other course—that is, south from Japan, and easterly through the islands of the Pacific, *against* winds
and currents, to the American shores.

In the equatorial regions of the Pacific, the prevailing winds and the currents, always flow from east
to west, or (in a plainer way) from the shores of the
two Americas towards Asia; the northern and southern
currents meeting at the equator off the Mexican coast,
and flowing together to the Indian Ocean, to part

again, and sweep around the North and South Pacific, as already described.

Between the Phillippines, the Japanese Islands, and the eastern coast of Asia, another current flows to the south, and into the Indian Ocean; a portion sometimes reaching the Peruvian current south of Australia, and running with it in its southern course.

This inner Asiatic current, if it may be so called, explains the total absence of Chinese wrecks in our northern regions, and at the same time accounts for the Chinese wrecks found in the Indian Ocean, and even at the Straits of Magellan, on the west coast of South America.

If we readily accept the views of many writers, the peopling of the Americas by the Asiatics was but natural and easy of accomplishment. If we examine history, facts and dates, we do not find the easy views advanced, sustained by them.

ASIATIC INFLUENCES IN PEOPLING AMERICA.

Grotius says: The Peruvians were a Chinese colony, and the Spaniards found, at the entry of the Pacific Ocean, on coming through the Straits of Magellan, the wrecks of Chinese vessels.

There are proofs, clear and certain, that Mango Capac, founder of the Peruvian race, was the son of Kublai Khan, the commander of this expedition, and that the ancestors of Montezuma, who were from Assam, arrived about the same time. Every custom described by their Spanish conquerors proves their Asiatic origin.

Again: The Hindoo, Chinese and Japanese annals all correspond in recording the fact that, about the year

1280, Genghis Khan, a great Mongol chief, whose name was a terror in Europe, at the same time invaded China with hordes of barbarians from Tartary, whom his descendants hold in subjection at the present time. Having accomplished this object, he fitted out an expedition consisting of 240,000 men in 400 ships, under command of Kublai Khan, one of his sons, for the purpose of conquering Japan. While this expedition was on the passage between the two countries, a violent storm arose, which destroyed a great part of the fleet, and drove many of the vessels on the coast of America.

(Cronise: Wealth of California.)

Some of these statements are hardly clear. The races from which the Montezumas sprung, were natives of Atzlan, a country forming at that time a small portion of northern South America, and extending into South Central America. In about 1180 A. D., a portion of this race emigrated to the valley of Mexico, forming the foundations from which the Aztecs sprung. If this statement be true, Kublai Kahn did not arrive in America until many years after. If the dates are correct, neither he or the people who are said to have reached America from Assam, about the same time, can be claimed as the founders of the Aztec race.

Probably if a thousand years or so were taken from the above dates, and time given for the great oceanic laws governing the currents of the Pacific, as well as the gradually extending ventures of a natural maritime people, like the Chinese and Japanese, we might account for a partial peopling, at least, of the Americas by the Asiatics.

Nor is it well, in this connection, to isolate ideas

and facts, and view the peopling of the Americas from
the Pacific standpoint alone, or to ignore the influence
of the great ocean currents of the Atlantic, or the
early maritime ventures of countries not on our side
of the world, and the bearing they have had on the
ethnology of America.

ISLAND RACES.

Among the islands of the Pacific, the lines separating races are very closely defined, and through what
would seem perfectly natural causes. In nearly every
case the peopling of the islands can be accounted for,
by supposing that their migratory habits were in ac-
cordance with the natural laws controlling the winds
and currents in these regions.

Closely following the migratory movements of the
human race, as an example, we may take the animal
kingdom. A north and south line can be drawn through
the Eastern Archipelago, where animals of the larger
growth cease to exist. Borneo, Sumatra, Java and
some of the other islands have the animal kingdom
of India and Asia well represented in the elephant,
lion, tiger, panther, rhinoceros, hippopotamus, ourang-
utang and monkey, with the reptilian and feathered
species of the larger kind, all partaking of the species
found on the main land of Asia. Of this latter country, it is believed that the islands named, at one time
formed a part.

Still another parallel, running north and south
and further to the east, may be drawn, where the
larger of the species named above, have never been
known to exist. Thus, the islands of New Zealand, Tasmania, Australia, New Guinea and others in the same

range, are entirely free from the animals enumerated, excepting the monkey tribes, and in Australia the kangaroo.

Another parallel can be traced, running north and south and still further east, through the island groups of the Society, Tongas, Fijis, Samoas, Marshalls, New Hebrides and the Carolines, where hardly any animal larger than the dog or rat, can be found native to the soil. These parallels are followed just as closely by the reptilian and feathered tribes.

The latter, whose migratory powers are well known all over the world, seem curiously to draw the species line of locality or habitation, as closely as those of the animal kingdom. In the Bird of Paradise we find a marked instance. Their native home is New Guinea, where as many as twenty of this species of birds may be found, and are hardly ever to be met with in any of the other island groups.

This follows, also, in nearly as strictly defined lines, with the inhabitants of Oceanica. The people of Borneo, Java, Sumatra and the Molluccas partake of the Malay, Hindoo and Chinese, being all, in a comparative sense, a maritime people.

At Australia this race element ceases altogether. The natives are bushmen, and root-diggers, with no knowledge of navigation; not canoe-builders, or fishermen, nor in any way resembling a people who "go down to the sea in ships." The same is true of the New Zealander and the Tasmanian. Yet, but a little to the north, on New Guinea, and in the Carolines, the natives have some knowledge of canoe-building, sailing and maritime ventures. So on through the Molluccas and Phillippines, into Japan, where the art of ship-building and navigation, as among the islanders of the Pacific, may be said to have been brought to comparative perfection.

East from Australia, in the Solomon Archipelago, and among the Marshall islanders, the Samoans, in fact, as far east as the island groups extend, north and south of the line, the Asiatic features are prominent. The inhabitants are expert canoe and boat builders, with considerable knowledge of navigation, making long voyages in their little crafts with lateen sails and outriggers to windward, and altogether perfectly at home on the water. These people, with the exception of the Fijis, and others of the wooly-headed type, have the features and many of the characteristics of the Chinese and Japanese—probably coming from those countries, making the grand circles of the ocean currents, with favoring winds, at very early periods.

The many wrecks of Japanese vessels found in the Northern Pacific, following the line of the ocean currents clear into the island groups, seems important evidence in favor of the above statement.

A like statement may be made of the maritime ventures of the Chinese, south of the equator, many traces of whose early settlements, habits and architecture are to be found in South America.

This would account for the absence of animal life of the larger kind on the easterly islands, as the length of the voyages, together with the small size of the shipping of the earlier periods, would make the carrying of animals almost an impossibility.

The prevailing winds follow the course of the currents through the equatorial regions of the Pacific from east to west. Assuming the movements of the ocean streams to be twenty-one miles per day, and

that favoring winds would add to the floating powers
of a boat or canoe fifteen miles a day additional, we
would have a favoring drift from east to west of
thirty-six miles per day. Thus we might assume, that
a journey of 1,000 miles per month could be made
without the aid of sails or oars. *Against* such favoring circumstances it does not seem possible for a people without the modern appliances of steam and sail,
to migrate.

Many traces of ruins of architecture, similar in
form to the pyramidal structures of the ancient Peruvians and Chilians, are to be found in some of the
islands, on Ascension particularly. Great blocks of
hewn granite are to be found, with other forms of
building stone, scattered over the ground in many
places, and lying under water in some of the harbors.
It was thought at one time that these had been transported from great distances, and that the geological
formations of the material were foreign to anything to
be found on the islands. Closer research, however,
revealed the quarries from which the stones had been
taken, located in the interior of the islands where such
ruins were discovered.*

This fact has spoiled many curious, mysterious
theories that were advanced in regard to the building
material, and leaves us but to account for the people
whose intelligence and skill, indicates their source to
be from countries foreign to these islands. From the
data (a review of which would but tire the reader) obtained on this subject, the race origin of many of the
islanders of Oceanica is clearly indicated to be Chinese
and Japanese.

*The stone implements, with the hieroglyphical writing and draw-
ings on the rocks, found on Pitcairn by the Bounty mutineers, may
help, some day, to trace the history of the ancient islanders

THE EQUATORIAL CURRENTS.

As my purpose has been throughout this work to present facts, untrammeled by personal opinion, for the consideration of the reader, I add a few notes below, taken from experiences and researches of others, that may modify or change altogether some of the ideas already advanced:

The famous volcanic eruption on the island of Krakatoa, just west of Java, a year since, startled the civilized portion of the world with the "blue" and "red" and other "strange sunsets and sunrisings" it caused. Just now, a year after date, Ponape is gathering up some of the products of that eruption; large beds of pumice-stone, in places, are covering the sea with its gray hue, as if an immense blanket were spread out. Months since, I saw an account of one of the harbors, near that eruption, filled with this material ten feet deep, and almost as compact as an ice-floe. The winds, and especially the currents, have taken some of that disgorged mass and floated it to our Ponape reefs. A remarkable fact about this is the continuity of an easterly or northeasterly set of the ocean's current near the line. No doubt masses of the ejected pumice will float along on the same current to the shores of South America, more than half way belting the earth. Our natives call it "sea-fruit," for they have no idea where or how it was gendered, but suppose the sea is the mother.

To some of the sandy coral islands lying in the track, it will be a very god-send. The material is gathered, crushed, and put on beds of taro as a fertilizer. Mere sand-beaches, or banks, furnish but little to fertilize vegetation.

But Krakatoa, or Krakatao, has other interests to
Ponape. The word is of two syllables—the first, the
specific name, and *tao* or *tau*, meaning "strait;" hence
the term means "Kraka of the strait." But *tao* or *tau*
is pure Ponapian, and here also means a strait—a passage of water. Java, then, and Ponape are blood-
related. Indeed, centuries and centuries since, at
least as far back as when Solomon was king, Java had
another kind of eruption, sending off here ever so
many of her vocables. But recently I counted more
than fifty of these, some of them names of places on
this island. These vocables, of course, took passage
with the Malay tongue. And now Java is sending
fields of pumice-stone. Some day those who are on
the east of her must send back or set afloat to her,
truths from God's Word.

(Rev. Edw. T. Doane, Ponape, Micronesia).

This would indicate an equatorial current flowing
from west to east, in an opposite direction to, and
between, the two great ocean currents of the Pacific.
The speed of the current would be about eight miles
per day, if we estimate the distance from the island of
Java to that of Ponape to be 3,000 miles.

Again—from Wallace, Muller, Dr. A. B. Meyer,
Schouw-Santvoort, Proc. Roy. Geo. Soc., 1881, and
Ency. Brit., vol. 15, I quote the following:

Long considered as an independent division of
mankind, the Malays are now more generally affiliated
to the Mongol stock—of which A. R. Wallace, De
Quatrefages and other eminent naturalists regard
them as a simple variety, more or less modified by
mixture with other elements. These considerations
also enable us to fix the true centre of dispersion of

the Malay race, rather in Mallacca than in Sumatra, contrary to the generally received opinion. If they are to be physically allied to the Mongol stock, it is obvious that the earliest migration must have been from High Asia, southward to the peninsula, and thence to Sumatra, possibly at a time when the island still formed a part of the mainland. The national traditions of a dispersion from Menangkabo or Palembang, in South Sumatra, must accordingly be understood to refer to later movements, and more especially to the diffusion of the civilized Malay peoples, who first acquired a really national development in Sumatra, in comparatively recent times. From this point they spread to the peninsula, to Borneo, Sooloo, and other parts of Malaysia, apparently since their conversion to Islam, although there is reason to believe that other waves of migration must have reached Further India, and especially Camboja, if not from the same region, at all events from Java, at much earlier dates. The impulse to these earlier movements must be attributed to the introduction of Indian culture through the Hindu and Buddhist missionaries, perhaps two or three centuries before the Christian era. During still more prehistoric times, various sections of the Malay and Indonesian stocks were diffused westward to Madagascar, where the Hovas, of undoubted Malay descent, still hold the political supremacy, and eastward to the Phillippines, Formosa, Micronesia and Polynesia. This astonishing expansion of the Malaysian peoples throughout the Oceanic area, is sufficiently attested by the diffusion of a common Malayo-Polynesian speech from Madagascar to Easter Island, and from Hawaii to New Zealand.

"TABOO."

One of the curious customs among the islanders of the South Sea, is the practice of that rite, so little understood by the traveler, who is not "native and to the manor born"—"taboo."

Tabu, Tapu, or Tambu, a Polynesian term, denoting an institution found everywhere, and always essentially the same, in the Polynesian Islands and in New Zealand. Its primary meanings seem to be exactly the same as those of the Hebrew *toebah*. This word, like the Greek *anathema*, the Latin *sacer* and the French *sacre*, and the corresponding and similar terms in most languages, has a double meaning—a good sense and a bad; it signifies, on the one hand, sacred, consecrated; on the other hand, accursed, abominable, unholy. It results, from a thing being held sacred, that certain acts are forbidden with reference to it, and from any act deemed abominable; that it is forbidden. A notion of prohibition thus attaches to the word tabu, and this is in many cases the most prominent notion connected with it. The term is often used substantially in the sense of a prohibition—a prohibitory commandment. If a burial ground has been consecrated, *it is tabu;* to fight in it, then, is sacrilegious and prohibited, and this also is tabu; moreover, those persons are tabu who have violated its sanctity by fighting in it, and they are loosely and popularly said to have *broken the tabu*. This example illustrates all the uses of the word. It has furnished to the English language the now familiar phrase of being "tabooed"—that is, forbidden.

(Chambers's Ency.)

The observance of the custom among the natives of many of the island groups is universal at all times

and places, and fortunately has been the means, not only
of protecting strangers from insult and injury, but the
preserving of life as well. In Micronesia, the ordinary
native can select a favorite cocoanut tree, banana plant,
or the hut in which he lives, and protect them from
the inroads of all comers, either by erecting a monument of loose stones, or laying them in a peculiar
manner in front of his dwelling, or by tying a banana,
palm or plantain leaf around the tree or plant, which
indicates that it is tabu. Thus we see that the exercise of the right is not confined to the chiefs or people
of high degree, but is in general use among the lower
orders. The women, unless wives or daughters of
chiefs, are not allowed to exercise the right; yet a man
may protect any of the sex from insult or injury, by
the observance of the forms required.

A SMALL TRIBUTE TO RELIGIOUS MISSIONS AND MIS-
SIONARIES.

The inception of religious missions dates far back
in the biblical ages. Their history, or the life and
works of a people who practice what they preach, and
convey the good they have acquired from religion, civilization and enlightenment, to those of the world less
fortunate in this respect, would fill volumes.

It is but little to praise the efforts of patient and
daring workers, pioneers of light, in distant, dangerous,
inhospitable lands, or speak of the many, rich and poor,
who contribute a portion of their effects to the good
cause—even to the widow's mite—and furnish the
sinews of war to a noble army of Christian workers,
the benefits of whose enlightening course through the
pagan world can hardly be overestimated.

With this in view, I take pleasure in citing a few
of the great benefits resulting from the works of Christian missions in modern times.

In China, Japan, India—in fact, in all parts of Asia,
Africa, the two Americas, and in Oceanica, we find
their churches and schools. Following closely in the
footsteps of adventurous missionaries, we see that boon
to mankind, the printing press. Used not alone in
the translation of the Bible and religious works, but,
as in Shanghai, where ten presses are in almost constant use, we find them printing works on science,
medicine, law, history, agriculture, school books, etc.,
and scattering them broadcast throughout the land.
Thousands of volumes, on one hundred and fifty different subjects, are printed and circulated among the
people. And all this but a tithe of the work accomplished among the pagans of other countries. Christian missionaries have translated the Bible, school books,
and hundreds of other instructive, useful works, into
over two hundred languages and dialects.

Many of them, in addition to their sacredotal
acquirements, are educated physicians as well. At the
principal stations of the mission world, medical dispen-
saries are to be found, whose drugs, skillfully used,
present an effective barrier to the spread of epidemical
diseases. Of late days it has become customary to
educate the women of the societies in medicine,
to whose ministering cares thousands of pagans owe a
healthful existence.

In one district in Africa, between Sierra Leone
and Gaboon, a distance of nearly 2,000 miles, twelve
Protestant societies have established missions. They
have something over 20,000 children being educated in
their schools, and many more adults, as members of

MISSION HOME—CAROLINE GROUP.

Christian churches. Under this influence the slave trade has altogether disappeared, where in former times it counted its victims at the rate of 20,000 a year.

Among the 5,000,000 inhabiting the island of Madagascar, 500,000 are members of Christian churches.

Among the islands of the Pacific, particularly those of Polynesia, Melanesia and Micronesia, the advancement and benefits are fully as marked. Some sixty or seventy years ago, sunk in the degrading depths of paganism, a great many of them cannibals, now number over 500 islands under the care of the missions.

Over twenty of their languages have been reduced to writing. Churches and schools adorn the land; the sound of the axe, saw and hammer, with the busy hum of manufactures, replace grim war and the hideous rites and yells of the man-eater.

In these islands it has been truly said that hundreds of native teachers and missionaries, who have themselves attended the feasts and joined in the revolting rites of the cannibals, may now be found successfully pointing the way, among their heathen brethren. The 200 churches and 1,400 schools in the Fiji Islands, the traditional home of the man-eater, will equally serve "to point a moral or adorn a tale" of missionary work.

Catholic and Protestant alike, are establishing religious stations in all parts of the pagan world, and with a friendly rivalry, that but adds strength and effectiveness to their efforts.

Many of the obstacles to be overcome by the missionary, particularly among the islands of the South Sea, are not the fierce intractable disposition of the

* 16

natives, but the barriers placed in the way by a low class of people, already referred to in this work. Beach-combers, wreckers and buccaneers, castaways from our civilization, have had more to do with the modern introduction of disease and degradation among the natives, than inherited paganism. The man who first taught them how to turn a pleasant, healthful drink, the sap of the cocoanut palm, into arrack, a vile brain-entangling rum, has introduced a degrading element more to be dreaded than pagan superstition.

CHAPTER XVI.

——

BIRTH, GROWTH AND DEATH OF ISLANDS.

——

Imprison'd fires in the close dungeons pent,
Roar to get loose, and struggle for a vent;
Eating their way, and undermining all,
'Till with a mighty burst, whole mountains fall.

ADDISON.

THAT great mystery of the Atlantic Ocean, sunken Atlantis, has formed the theme of tongue and pen for ages. Veiled in tradition and romance, little has been ventured in the way of a truthful explanation, of the fate of the great island and her people.

Yet in plain view, and without the garb of fiction, we have the birth and death of islands in almost constant operation in the Pacific, as well as in other parts of the globe. In this connection, I quote from a recent publication:

Geographers complain that soon there will be no more worlds for them to conquer, and the Danes have ever since the loss of the Duchies, looked forward with doleful forebodings to the time when their country will be still further shorn of its fair proportions. Nature is, however, bountiful, and now, by throwing up a new island off the shores of Iceland, it has added

in an appreciable degree to the territories of King
Christian, and to the regions which still await the ex-
ploration of the traveler.   It is true, the new land is
only a volcanic cone, and as it was the result of subterranean fire, may, like so many of its predecessors,
born of the throes of mother earth, sink again into
the ocean from which it sprang.

At various times, especially after some severe
disturbance of Hekla, similar islands have shown
themselves above the waves, but generally, with the
exception of Nyoe, which was thrown up last century, have been worn away by the action of the surf,
before geologists could accurately examine the volcanic scoriæ and ashes of which they were composed.
In 1811 Captain Tillard, of H. M. S. *Sabrina*, witnessed such an islet arise during a volcanic outburst
in the Azores, and proudly named it after his ship.
But when he returned a few weeks later, to survey
and annex his acquisition, not a trace of Sabrina
Island was visible.   The sea had reclaimed it.   In the
volcanic region of the Mediterranean several similar
births of land have been recorded by ancient and
modern writers.   But the most notorious of them
was Graham Island, which arose in the year 1831,
some thirty miles off the southwest coast of Sicily.
For a few weeks much ink was shed over it, and at
one time it was feared that gunpowder would be
burnt in the assertion of the angry claims which were
made for the wretched 2,300 yards of Ætnaic cinders.
The names of Sciacca, Julia, Hotham, Graham and
Corrao were suggestively given to it by the fiery
mariners who cruised around it, ready to land and
hoist their countries flags the moment the scoriæ
cooled.   But before Europe was embroiled in war

about it, Graham Island vanished, and so settled the dispute in its own simple way.

After the destruction of Krakatoa by the great Javan earthquake of 1883, twenty-one new islands appeared in the Sunda Straits, and only last year, one hitherto unknown, rose above the sea off the shores of Alaska.

In all these cases, volcanic action has been the ostensible cause of the formation of these specks in the ocean, But in 1871 Captains Luzen and Mack discovered to the north of Nova Zembla, a group of islets just above the sea, on the very spot where, in 1854, William Barrant had found soundings. On the two largest, which were named Brown and Hellwalld's Islands, tropical fruits were picked up, tossed hither by the northern extension of the Gulf Stream. Hence the group was named the Gulf Stream Islands, and as the land in this portion of the Polar basin is undergoing a slow secular elevation, just as in other places it is sinking, in the course of a century or two the Arctic navigator may find in that direction something worthy of a flag and an entry on his chart.

From the latest date at hand, the islands formed in the Straits of Sunda, alluded to in the above article, have disappeared in the sea, and smooth navigable waters roll above their tombs.

VOLCANIC AND EARTHQUAKE LORE.

A small island lying off the northeast coast of Sumbawa, named Gunong Api, must here be mentioned, because it contains a volcano, and forms a part of that "belt of fire" to which we have adverted as one of the most remarkable physical features of the Indian Archipelago.

It is recorded that the inhabitants of Java, when the eruption began (on the above island), mistook the explosion for discharges of artillery, and at Jayokarta, a distance of 480 miles, a force of soldiers was hastily dispatched to the relief of a neighboring port that was supposed to have been attacked by an enemy. At Surabaya, gun-boats were ordered off to the relief of ships which were defending themselves, it was thought, against pirates in the Madura Strait; while at two places on the coast, boats put off to the assistance of supposed ships in distress. For five days these reports continued, and on the fifth the sky over the eastern part of Java grew dark with ashy showers, so that the sea was invisible. According to Mr. Crawford, the sky at Surabaya did not become as clear for several months, as it usually is in the southeast monsoons.

Eastward, the din of the explosions reached the island of Ternate, near Gilolo, a distance of 720 geographical miles, and so distinctly was it heard that "the resident sent out a boat to look for the ship which was supposed to have been firing signals." Westward, it was heard at Moko-moko, near Bencoolen, or 970 geographical miles.

Dr. Junghuhn thinks that within a circle described by a radius of 210 miles, the average depth of the ashes was at least *two feet*, a circumstance which will enable the reader to form some idea of the tremendous character of the eruption. The mountain, in fact, must have ejected several times its own mass, and yet no subsidence has been observed in the adjoining area, and apparently the only change is, that during the outbreak, Tamboro lost two-thirds of its previous height.

The Rajah of Sangir, a village about fourteen miles southeast of the volcano, was an eye-witness of the eruption, and thus describes it:

About 7 P. M., on the 10th of April (1815), three distinct columns of flame burst forth near the summit of the mountain, all of them apparently within the verge of the crater: and after ascending, separately, to a very great height, united their tops in the air in a troubled, confused manner. In a short time the whole mountain next to Sangir appeared like a mass of liquid fire, extending itself in every direction. The fire and columns of flame continued to rage with unabated fury until the darkness, caused by the quantity of falling matter, obscured it about 8 P. M. Stones at this time fell very thick at Sangir, some of them as large as a man's two fists, but generally not exceeding the size of walnuts.

Between 9 and 10 P. M. showers of ashes began to fall, and soon afterwards a violent whirlwind ensued, which overthrew nearly every house in the village of Sangir, carrying along with it, their lighter portions and thatched roofs. In that part of the district of Sangir, adjoining the volcano, its effects were much more severe; it tore up by the roots the largest trees, and whirling them in the air, dashed them around in the wildest confusion, along with men, houses, cattle, and whatever else came within the range of its fury. The sea rose nearly twelve feet higher than it had ever been known before, and completely destroyed the only small spots of rice lands in Sangir, sweeping away houses and everything within its reach.

The captain of a ship dispatched from Macassar, to the scene of this awful phenomenon, stated, that as he approached the coast, he passed through great quantities of pumice stone floating on the sea, which had at first the appearance of shoals, so that he was deceived into sending a boat to examine one, which at the distance of a mile, he supposed to be a dry sand-bank, *upwards of three miles in length*, with black rocks projecting above it here and there.

Mr. Bickmore speaks of seeing the same kind of stones floating over the sea, when approaching (in April, 1865) the Strait of Sunda. He adds: Besides the quantities of this porous, foam-like lava that are thrown directly into the sea by such eruptions, great quantities remain on the declivities of the volcano and in the surrounding mountains, much of which is con-veyed by the rivers, during the rainy season, to the ocean.

(Bickmore: Travels in the Eastern Archipelago.)

VOLCANIC FIRE-BELT OF THE WESTERN HEMISPHERE.

Humboldt gives a list of the volcanoes of the world, calculated many years ago. It therefore may be accepted as under-estimated, as there are some 900 volcanoes, extinct and active, to be found in the Eastern Archipelago alone.

Europe	7,	with	4	active.
Atlantic Islands	14,	"	8	"
Africa	3,	"	1	"
Continental Asia	25,	"	15	"
Asiatic Islands	189,	"	110	"
Indian Ocean	9,	"	5	"
South Sea	40,	"	26	"
North and South America	120,	"	56	"
	407,	"	225	"

As will be seen by the map accompanying this work, the volcanic fire-belt very nearly surrounds and outlines the western hemisphere. At Mount Erebus, but a few hundred miles from the South Pole, we see one of Nature's grandest outbursts—one of the world's greatest volcanoes in ceaseless eruption. With its lurid glare reflected back in a hundred ways by the icy mirrors of frozen seas, and the prismatic colorings of towering icebergs, it forms a spectacle too grand for description. Based and capped in the regions of perpetual ice and snow, its fiery peak, 13,000 feet, reaching up in the clouds, is a beacon light in an unknown, untrodden land.

THROUGH SOUTH AMERICA.

From this source we shall trace the volcanic, eruptic fire-belt. Making its way north, the great subterranean fire-stream — one branch of which passes under the South Shetland Islands, and on under the restless Atlantic; the other passes through Terra del Fuego, and across the Straits of Magellan into South America. Here the fiery current forces its resistless way under the towering peaks of the Chilean Andes, breaking out at the volcanic peaks of Acacagua, Hulliaciaca, Villarica, San Jose, Peteroa, Antuco, Hamatua, Chillan, Calbuco, Corcovado, Osomo and Zandeles. Through Bolivia, appearing in the volcanoes of Isluya, and Sajama, whose peaks tower 22,350 feet above the sea, and on into Peru, breaking out in angry flames in Arequipa, from the towering peaks of Mesta, Chacarni. Pan de Azucar, burying the cities of Arequipa and Orite, Tultapace and Ubinos, in burning lava and ashes, in the sixteenth century. And again, at Cotopaxi, 19,500 feet above the sea, boiling over and forcing its fiery way out of a height of 17,000 feet at

Sangaii, still in Peru, pouring out sulphurous smoke, ashes, cinders and lava, the flames lighting up the country around for one hundred and fifty years past. Hugging the Pacific shores, along into Ecuador, where the great extinct crater of Chimborazo lies, while a branch of the stream, now extinct, makes off to the west some six hundred miles or more, and burst out in the Galapagos Islands, whose numerous extinct craters, nearly two thousand in number, give evidence of a severe eruption in past ages.

CENTRAL AMERICA AND MEXICO.

From Ecuador, the current flows on through New Granada, Guatemala, Central America and San Salvador. The current through these latter countries seems to be in a quiescent state, as, although abundant evidences of its eruptic forces can be traced in the past, there are no active volcanoes in existence in those countries at the present time.

Still onward pursuing its northerly course, to break out again in Mexico, in Anahuac and in Michiochan, in the volcanoes of Tuxtla, Orizaba, Popocatapetl, Isztachuatl, Toluca, Jornillo, and in Colima, in Zapotai, Tancitari and Soconusco. These are nearly all in an inactive state at present, if we except a little smoke and sulphurous vapors emitted from some of the craters.

Tuxtla, though (in the State of Vera Cruz), emits a flame day and night, lighting up the heavens with a glare that may be seen far away at sea.

The current branches here again, one stream making its way due west, under the sea, for over 2,500 miles, to appear again in those majestic volcanic outbursts of Kilauea and Mauna Loa, in the Sandwich Islands.

PACIFIC COAST.

The other stream pursues a peaceful course on through North America, following the line of the Pacific shore, on through California, Oregon, Washington Territory and British America, into Alaska. Through these countries, the flow of the fiery channel below may be traced by the evidences, not only of extinct volcanoes, but of the vast overflow of lava and volcanic tufa, to be found all along the route named.

Of Mount Hood, Shasta, Mount St. Helena, and some others of lesser note, there is little to be said. Their peaks, rising from eleven to fourteen thousand feet, have no doubt formed vents for the restless fluid beneath. The geysers, hot springs and mud ebulli- tions, found all along the Pacific coast, owe their exis- tence and activity to the yet unsubdued fires of the volcanic belt.

THROUGH THE ISLANDS.

Breaking out again at Mount St. Elias, in Alaska, in fitful outbursts, and but lately on one of the islands of the Aleutian chain, we see the mighty forces of the fire-stream still at work.

Crossing from Alaska to Kamptchatka, through the Aleutian Islands, and touching the southern portion of the latter country, the eruptic current turns south- by-west, and flows on through the Kurile Islands, and through the main groups of the island empire of Japan, whose uneasy foundations are truly said to be rocked in the cradle of the deep.

Still onward, pursuing its southerly course, through the Phillipine and Mollucca Islands, often shaking them to their centers with its angry forces, the fire-stream makes its way, touching the northwestern portion of Celebes on the one hand, and missing its great island neighbor, Borneo, on the other; it bursts forth in terrible and oft-recurring eruptions in ill-fated Java. Here again the current divides, one sweeping to the north and west, through Sumatra, and away into the Bay of Bengal; the other turns at a point further north, from the Molluccas, and flows east-by-south, barely touching New Guinea, through New Ireland and New Britain, under the Solomon Archipelago; then again to the south it pursues its fiery way, through the New Hebrides, into New Zealand; while another, evidently smaller stream, branches just north of the Hebrides, flowing south-by-west, touching the southeastern coast of Australia, and apparently terminating at the island group of Tasmania or Van Diemen's Land.

As far as known, there are sixty-five volcanoes in Alaska, ten of them being active, with one or two more in the Aleutian Isles. In the New Hebrides, on the island of Tanna, a volcanic peak still forms one of the beacon lights of the South Sea, to be rivaled sometimes by its fiery neighbor, Tongariro, in New Zealand.

THEORY OF VOLCANOES AND EARTHQUAKES.

Many theories have been advanced by scientists, to explain earthquake and volcanic action; though that advanced by Darwin, from observations in nearly all parts of the world, is generally accepted. It is believed that the crust of the earth, slowly cooling from its once liquid mass, has now formed a crust of from ten to twenty-five miles in thickness, and still holds within this great covering or shell, a molten mass of subterranean fires, and that volcanic outbursts occur only within certain lines—probably those where the earth's shell is thinnest. It has been noted that the eruptions are more frequent—in fact, take place altogether—where the earth's surface is raising, being pushed up by the mighty forces within its shell. Eruptions never occur in lines where the crust is sinking or undergoing a depression, on account, no doubt, of its immense weight, thickness, and the additional strength it has acquired from cooling. The theory, sometimes advanced, of the cracking and rending of the cooling shell, and allowing the waters of the seas to penetrate to the subterranean fires, with the consequent eruptive forces created by steam, would more than explain the earthquake phenomena. That the earth's shell would close again, after admitting just enough water to give an exhibition, such as we see in volcanic outbursts, is very doubtful. It is more than likely that the two elements, fire and water, coming together in the manner described, would rend the world from pole to pole, and leave us little but the theory to contemplate, if that.

The cause of earthquakes has already received considerable attention, particularly those continually occurring all over the world, unaccompanied by volcanoes. Earthquakes with the wave motion, attended by an indescribable rumbling roar, are judged to be the offspring of restless subterranean fires; while others, with the quick-recurring, nervous shocks, and of which California furnishes many examples, are accounted for by electrical movements taking place between the great elements, earth, air and water. Again, these apparent electric shocks are explained, by assuming the crust of the earth to be opening in cracks and fissures, and that the formations are slipping, one by the other, giving such a motion to the surface, as one may experience by forcing the moistened finger over a surface of glass.

CHAPTER XVII.

COMMERCE, AND INTEROCEANIC CANALS.

> A storm-cloud, lurid with lightning,
> And a cry of lamentation,
> Repeated and again repeated,
> Deep and loud
> As the reverberation
> Of cloud answering unto cloud,
> Swells and rolls away in the distance,
> As if the sheeted
> Lightning retreated,
> Baffled and thwarted by the winds' resistance.
>
> LONGFELLOW (*Christus*).

REVERTING again to the commercial interests locked up in a great portion of the island world, and which but awaits the key of American energy and enterprise to open and develop, the reader may find the following chapter entertaining, by taking a general glance with me at some of the interests likely to affect the commerce and industries of America.

Professor Hanks says: As the domestic, and the other material interests of California, have prospered and expanded, so also has the commerce of the country grown into large proportions. With an import trade second only to that of New York, San Francisco has such virgin fields to occupy, as open not to her great eastern rival. To her the trade of

Australia and the Orient, including Eastern Siberia and the islands of the Pacific, geographically as well as commercially, belongs; time, freights, interest and insurance all being in her favor, as against every other port in the world.

Although the trade of San Francisco, which may be said to represent largely that of the State, has suffered in some of its departments, through the construction of two additional transcontinental railroads—the one to the north, and the other to the south, of the more central route—it still continues large, and has even increased in the aggregate, since the completion of these lateral lines, indicating that this trade is not likely to be seriously crippled by this or other interfering causes.

The value of the merchandise and treasure shipped from San Francisco in 1883, amounted to $105,000,000, of which $46,000,000 were consigned to foreign countries. Of these exports, $60,000,000 went by sea, and $45,000,000 by rail. The imports from foreign countries amounted, meantime, to $40,000,000; the following staples, among other leading articles, having been imported in the amounts here mentioned: Sugar, 133,914,154 pounds; rice, 58,315,750 pounds; tea, 20,960,248 pounds; and coffee, 17,444,777 pounds. The receipts of lumber at this port amounted, for the year, to 276,772,469 feet, valued at $5,000,000; receipts of Federal revenue, $12,558,305.

The innumerable plants and trees in the Pacific, whose bark, pith and fiber, now worked in a crude way among the natives, into paper, cloth and fibrous manufactures, could be built up into a large profitable trade under more civilized rule. The pulp could be pressed, dried, and shipped, say to San Francisco, where a paper, rivaling the celebrated linen products of that article, manufactured in Europe, could easily be produced.

The black walnut, Spanish cedar, toa, tomano and prima vera, the rosewood, dye-woods and mahogany, growing so profusely in the island world, the satin, sandal and camphor trees, back up the assertion that immense commercial transactions with the Pacific Islands are in the near future.

The cordage interests might be developed in much the same way, by importing the many forms of the raw material, which nature produces in the Pacific Islands, and manufacturing them into the various articles required in our advanced civilization. As the reader is already familiar with many of the natural and cultivated products of the island world, a repetition here would prove uninteresting. The return trade of America with the islands is growing rapidly from year to year. Our breadstuffs, dry goods, canned goods, clothing, hardware, machinery, lumber, etc., now forming a considerable part of the shipping lists of commodities being forwarded to the Pacific Islands, are growing in quantity and value from year to year.

So vast and valuable are the commercial interests of the islands of the Pacific, that estimated on the actual product of the Hawaiian group alone, and this on their exports only, and that to one port, San Francisco, that any estimate on the commercial possibilities of the future, would but excite the doubt and ridicule of the skeptical reader,

In round numbers, the export of the above islands to the port named, is say, 100,000 tons per annum. In comparison with the area of the available lands located in the Pacific, the above group would constitute

* 17

but the 760th part; or the whole, would export some 76,000,000 tons per year to San Francisco alone. To transport this tonnage, 15,200 1,000-ton steam or sailing vessels would be required, making five round trips per year. Assuming that San Francisco is but a distributing point, and that, too, by rail, it would require 13,800 freight trains, carrying net 300 tons per train, or 690 trains per day, or a train would have to leave our city about every two minutes, day and night. Allowing that the trains would require twenty days to make the round trip, the above number, 13,800, would be required.

If we take but twenty per cent. of the above, we would yet have a practical trade so vast that a city of a million or more inhabitants would naturally be required to take care of it.

Assuming again that the value of the exports of San Francisco to the Hawaiian group would compare as favorably with all other portions of the island world of the Pacific, the value would be something like $2,432,000,000 per annum, over three times the value of the annual exports of the United States.

PANAMA CANAL.

The proposition to connect the Atlantic and Pacific Oceans by means of a canal, the work on which is now under, it is to be hoped, successful progress at Panama, will add greatly to the world's interest in the Pacific Islands. Of the many projects to connect the two oceans, if we add Captain Ead's ship railway, and similar schemes, the canal at Panama is about the *fifty-fourth.* The subjoined memorandum statement of the three most prominent undertakings, and for which I am indebted to the valuable writings of Captain W. L. Merry, gives a comparative idea, not only of their magnitude, but of the practical results, that will be derived after the completion of either of the proposed routes.

MEMORANDUM OF PANAMA CANAL.

Length of Panama railroad, 47.5 miles; length of United States Panama lock canal, 41.7 miles; engineer's estimate of cost of United States lock canal, including 20 per cent. contingency, $94,511;360; engineer's estimate of French sea level canal, including 10 per cent. contingency, $168,000,000.

Mercantile estimate of *probable* cost of French low tide level canal, San Francisco Board of Trade, $300,000,000.

Summit level of Panama canal survey, 295.7 feet; engineer's estimate of time for construction, 8 years.

To judge of the character of this work, the following estimate from the French survey is given herewith:

Length of dam, 5,000 feet; height above bed of the Chagres, 130 feet; height above canal level, 172 feet; height above canal bottom, 199 feet; estimated cost, 10 per cent. contingency, $20,000,000.

It will be noted that the bottom of the canal passes *in front* of the dam, seventy feet below the river bed, and that the Chagres River is *wiped out of existence* between the canal and the Atlantic. When the enormous rainfall, the violent freshets, and the large amount of sediment and floatage, brought down by floods, are considered, one begins to realize the enormous diffi-

culties of the project, the doubtful results of the attempt, and the impossibility of estimating additional
cost, which may be caused by contingencies liable to
occur. Presuming its completion, will this dam not be
a standing menace to the canal, passing in modest
silence two hundred feet below its top? What will be
the result of a moderate earthquake shock, or of seep-
age during the rainy season? Thus obliterating the
Chagres, the canal passes on into the Culebra division,
cutting through an elevation a few inches less than
three hundred feet—of course, with an immensely
increased excavation, as compared with the United
States survey, but encountering otherwise no formidable engineering obstacles—and finally reaching the
Pacific through the valley of the little Rio Grande,
about six miles west of the city of Panama, and there
meeting deep water about four miles outside the high-
water mark. The mean sea-level of both oceans is
now known to be the same; but, while at Aspinwall
the tide ebbs and flows from one and a half to two feet,
at Panama the tidal movement is eighteen to twenty-
six feet.

The American, as well as the French survey, overcome the difficulty by placing a tidal lock at the Pacific
end of the canal, which completely controls the question. Such is the French survey for a sea-level Panama canal.

NICARAGUA CANAL

Of the route of the Nicaragua canal, the following
memorandum will serve for a brief explanation:

Total length of interoceanic navigation, 173.57
miles; canal from San Juan del Norte to San Carlos dam, 35.90 miles; slack water navigation from

San Carlos dam to lake junction, 63.90 miles; lake
navigation from lake junction to lake end of Pacific
division of canal, 56.50 miles: extreme summit level
between Pacific and Atlantic Oceans, 150 feet; total
length of canal to be constructed, 53.15 miles; engineer's estimate of cost, $52,577,718; engineer's
estimate of time for construction, five years.

Mercantile estimate of *possible* cost by San Francisco Board of Trade, $100,000,000.

Surface of Lake Nicaragua is 107 feet 10 inches
above sea level. The Lake is 110 miles long and
about 35 miles wide, with average depth of water of
9 to 15 fathoms.

The Pacific division of the canal is 17½ miles
long, from Lajas on the lake to the Pacific seaport of
Brito.

THE EADS TEHUANTEPEC SHIP RAILWAY.

The survey for this interoceanic project has not
been made, and it is accordingly impossible to give an
accurate description of the line, or its exact length.
The Tehuantepec Isthmus United States canal survey
is 144 miles long, to which is to be added about 28
miles of river navigation, making a total of 172 miles;
and former surveys for railway and canal service, have
found the lowest practicable summit at 754 feet. The
canal project for this route was abandoned, because of
the high summit, necessitating a large number of locks,
with a scant water supply, while a tide-level canal is
impossible at any admissible cost. For a ship railway,
it offers advantages over any American isthmus, and
an ordinary railway is now being constructed there by
an American company. The Coatzacoalcos River is a
stream of respectable magnitude, running northerly

across the northern slope of the isthmus, with twelve to thirteen feet of water on its bar, which it is proposed to deepen sufficiently to admit the largest ships, which can ascend the river about twenty-five miles—how far, before arriving at the Atlantic end of the proposed railway, I presume Mr. Eads himself has not decided. There are no formidable obstacles in the way of building an ordinary railroad across the isthmus, beyond the heavy cuts and fills usually found in a country of that character; and the railroad finds its Pacific terminus at Salina Cruz, near Ventosa, at the head of the Gulf of Tehuantepec, where a port must be constructed. Probably Captain Eads can improve the Coatzacoalcos River for heavy navigation, 25 to 28 miles, and his railroad will be about 123 miles long. He estimates the cost at $75,000,000. It has been my purpose to avoid a discussion of the merits of the three routes here described, but it will be impossible to do so in the case of this project, if the reader is to acquire an intelligent idea of it. My high respect for the ability of Captain Eads, my esteem for him, founded on a slight personal acquaintance, and the fact that I can lay claim to no technical knowledge of civil engineering, are good reasons for approaching this subject with deference, and I must regard myself as merely a student of the project.

Captain Eads takes the ship out of water by a submerged inclined track, on which the cradle is run deep enough to allow the ship to be placed upon it, properly lined and blocked, after which a stationary engine hauls cradle and ship out of water to the railroad proper, where four "Mogul" locomotives are placed ahead of it, on a twelve-rail track, which haul ship and cradle to the other end of the track, where, by a reverse process, the ship is again placed in the water. Of course, there must be a cradle in use for each ship being transported simultaneously. The grades are overcome by *tipping-tables*, and the curves by *turn-tables*—as can readily be imagined, of gigantic size. How many of these he will need, cannot be known until surveys are completed; but I fear the Tehuantepec Isthmus will give him many grades and curves. He at first estimated the cost of such a railway at half the cost of a ship canal, but his present idea is, that it will cost $75,000,000, which at once detracts from his scheme the principal merit heretofore claimed for it, which was comparatively small cost; for there is every prospect that the Nicaragua Canal can be constructed for a like amount; and, while the depreciation and wear and tear of his railway, subjected to the action of a tropical climate, will necessarily be great, a ship canal improves with age—considerations of no little importance.

That Captain Eads can construct a ship railway across Tehuantepec, there is little doubt; that he can so construct it, as to meet all the requirements of the case, is another consideration. His mechanical appliances for overcoming the objections I was able to point out to him, appeared complicated, while the engineering obstacles of curves, grades, etc., his intimate knowledge of his profession had already indicated methods placing them under his control. He was willing to handle a loaded ship as carefully as I demanded, while it was my object, not to allow previous prejudices to affect my judgment of the merits of the scheme. In one respect, however, I fear, he has underrated the difficulty of his project. I doubt if, at Tehuantepec, or on any tropical American isthmus, he can find a

foundation for such a road as he wishes to build. The
"cuts" may support it, but the "fills" may fail to do so.
The success of the scheme depends on extreme rigid-
ity of road and cradle, and if, in tropical countries,
foundations are always troubling railroad engineers
under ordinary tracks, what are we to expect, under a
weight of fifteen or twenty thousand tons, concentra-
ted within the limits of the cradle carrying the loaded
ship? Captain Eads is one of the greatest living engineers, and if capitalists will furnish funds, he may build
his railway; but, unless it is cheaper than a canal,
what advantage does it offer? Why try an experiment,
when a certainty offers the same results? However,
in the absence of a survey with instruments of preci-
sion, it is probably unfair to discuss the project at all,
and I dismiss it, with great respect for the ability and
resources of the illustrious projector.

COMMERCIAL RESULTS ANTICIPATED.

That an American interoceanic canal will effect
great changes in the world's commerce, none can
doubt; but what little I shall have to say on this
branch of the subject, will refer to the effect it will
have upon American commercial interests generally,
and especially upon the interests of the Pacific coast
of our country—commercial, agricultural and social.
A project which brings this coast nearly nine thousand
miles nearer our Atlantic sea-board, and the great
marts of Europe, cannot fail to work great changes in
our commercial position. The inhabitants of the Pacific
coast must, for a long period, continue rather a producing, than a manufacturing people; and what manufacturing we are able to accomplish, will be from our

own products. The saving in time, insurance, depreciation and freights, applicable to Oregon and California, alone, will amount in ten years to the cost of the Nicaragua Canal. The saving above named, applied to this year's Oregon and California wheat crop, can be placed, with sober truth, at fully eight million dollars! When our wool, wine, and other growing industries are considered, it will easily be seen, that the producers of our coast should strain every nerve to insure the success of an interoceanic canal.

Nor, as might at first sight appear, will the canal injure our local railroads. While it would undoubtedly at first deprive them of the through freights, or force upon them a reduction which would be a great benefit to our State, in a short time after its completion their local traffic would surpass all the through traffic they can hope to control, and, with our other interests, they can reap the benefit of our rapidly increasing development, carrying all the products of our soil to tide-water, and securing a greatly increased passenger traffic. Meanwhile they have probably six years during the period of construction to accommodate themselves to the change.

The completion of the canal will make San Francisco the distributing point for the products of China, Japan and Central America, as far east as the Missouri, for it will then be to the interest of our railroads to secure this distribution rather than allow it to be made westward from Atlantic seaboard cities after reaching them through the canal. A rapid development of the Central American States and west Mexican coast would ensue, and those markets would increase their demand upon us for the commodities we are already sending there in limited quantity. Our merchant steam marine would rapidly increase, for the commerce between our eastern seaboard and our west coast being coastwise, and shut out from European competition, we should need a large steam tonnage under American colors to carry our freights eastward, while they would also compete with foreign steamers for European freights. It will be a glorious day for our State when San Francisco wharves will be crowded with four and five thousand ton screw steamers flying our flag and loading with our products, and with the completion of the canal this day will surely come. Cheap communication with Europe will bring to us desirable European immigration to settle up our lands and displace the unassimulative Chinese who are trying to crowd in upon us. Shall we not tend to keep them out by filling the places they would occupy with a class of immigrants that can be Americanized? An intelligent mind investigating this subject finds the grand results unfolding themselves until an interoceanic canal appears the greatest boon our coast can ask for, and to the names that are associated therewith, their country and the world will accord undying luster.

POLITICAL CONSIDERATIONS OF THE CANAL QUESTION.

Primarily, it would appear that it matters little who constructs a canal if our country is accorded the unrestricted use of it, in common with other nations. A further inquiry, however, must satisfy us that if we do not build this work we must acquire a controlling interest therein. We cannot afford so important a link in our coastwise communication to remain in the hands of any European organization, which would naturally consult foreign interests rather than our own. The Central American republics are now friendly to us, although sparsely inhabited and without development. The company constructing and managing an interoceanic canal would soon wield an influence paramount to the local government, and the policy of the latter might become subservient thereto and inimical to us.

During the existence of the Panama railroad it has been deemed a necessity for our government to keep armed forces almost constantly at both ends of the transit, and these forces have often been landed and kept ashore indefinitely for the protection of life and property. If this has been the case with a railroad managed by permanent *employees* and with a small native population, what may we expect when five to ten thousand laborers of various nationalities are congregated there, subject to a lax police control, suffering from malarial fevers, discontented, mutinous, and with a free supply of *aguardiente?* Add thereto a greatly increased native population, and we have all the elements needing military power to control them in emergencies.

When Count de Lessep's company have purchased the Panama railroad, which they have agreed to do as a preliminary step, we no longer have large American interests to protect there. It will be natural, and indeed necessary, for him to call upon the French Government to protect the enterprise, as we have protected the railroad company on many occasions. The French Government, both during and after construction, will find it necessary to station armed forces at both ends and on the line of the canal. After landing these forces a few times, what

more natural than that they should see the advantage and economy of having these troops in barracks
on shore—always within call? If it is claimed that
the French Government accepts no responsibility in
this connection, why has it already appointed an
official agent to oversee the initiation of the work?
If, at the end of our late internal war, our Government deemed it necessary to request the French to
promptly leave Mexico—merely contiguous territory—how much more important that they should
not be placed in a position completely controlling our
coastwise commerce, and establishing, first, their influence, then their power, and lastly, if we are quiescent,
their flag on the American Isthmus! Are the American people prepared for this? The late William H.
Seward, than whom no brighter intellect ever graced
American history, was wont to say that the Pacific
Ocean is to be the scene of man's greatest achieve-
ments. Are we prepared to have the key thereto in
foreign hands? Every American heart will say nay,
and honor the patriotism of President Hayes and
General Grant when they foresee these results and
point them out to their countrymen.

Nor is a large army and navy a necessity in the
maintenance of the Monroe doctrine; on the contrary, both *would become* a necessity were it to be dis-
regarded. The United States have a moral prestige
sufficient to create a respect for our rights and interests, and it is far better to meet attempted European domination on this continent, with a decisive
negative *now*, than to object thereto after it has passed
the initiative. It matters little where the capital comes
from to construct an interoceanic canal, but a due
respect for our national and traditional policy, as well

as for our national pride, should indicate the propriety
of its accomplishment through an American organization; and it is a poor compliment to our discernment
that we are to be kept quiescent by an "*American
Branch*," which can any day be voted out of existence at the headquarters of the Panama Canal Company in Paris! Americans will not fail to appreciate the
words of one who has proved himself worthy of their
patriotic regard: "I commend an American canal, on
American soil, to the American people!"

CHAPTER XVIII.

CURRENTS, WINDS, RAINS AND STORMS OF THE PACIFIC.

Bursts as a wave that from the clouds impends,
And swell'd with tempests on the ship descends;
White are the decks with foam; the winds aloud
Howl o'er the masts, and sing through every shroud:
Pale, trembling, tired, the sailors freeze with fears
And instant death on every wave appears.

POPE'S (*Homer's Iliad.*)

I DO not design, in this chapter, to more than glance, with the reader, at the broad expanse of waters, the majestic Pacific Ocean, and, in a general way, view its rains, storms and currents. Many men, wise in experience and intellectual acquirements, have already given these interesting subjects their careful attention; our hydrographic offices, and the shelves of our more advanced libraries, teem with the rich results of intellect and experience. The general flow of the great currents, with the rise and fall of the tides, and the natural laws controlling the winds and storms, on the great waste of waters of the mighty sea, are clearly depicted on charts, while elaborate data fill our nauti-

cal almanacs, sailing directions, and kindred works.
Yet so vast is the Pacific, that local influences are
occurring in many forms, and in many places, and all
acting without one influencing the other. Thus, if we
could be transported, as fast as the mind can travel,
from the Arctic to the Antarctic Oceans, or from the
Bay of Panama to the Bay of Bengal, or circle among
the intermediate latitudes or longitudes, all the climates of the world would be experienced, with their
varied physical influences, taking place at hundreds of
different localities, at about the same period of time.
So that none but the grander movements, like the
"Black Stream" in the North, and the Peruvian current in the South Pacific, and the main movements of
the equatorial currents, flowing both east and west,
with the ceaseless ebb and flow of the tides, are all
that can be contemplated with anything like certainty.
Maury, in his "Physical Geography of the Sea," gives
many examples of the variability of ocean currents.
He says, speaking of the Pacific:

There are also, about the equator, in this ocean,
some curious currents, which I have called the "Doldrum currents" of the Pacific, but which I do not understand, and as to which, observations are not sufficient
yet, to afford the proper explanation or description.
There are many of them, some of which, at times, run
with great force. On a voyage from the Society to
the Sandwich Islands, I encountered one running at the
rate of ninety-six miles a day. These currents are
generally found setting to the west. They are often,
but not always, encountered in the equatorial doldrums
on the voyage between the Society and the Sandwich
Islands.

In Captain Pichou's abstract log of the French

corvette *L'Eurydice*, from Honolulu to Tahiti, in August, 1857, a doldrum current is recorded at seventy-nine miles a day, west-by-north. He encountered it between 1 deg. north and 4 deg. south, where it was three hundred miles broad. On the voyage to Honolulu, in July of the same year, he experienced no such current, but in 6 deg. north, he encountered one of thirty-six miles, setting southeast, or nearly in the opposite direction. This current does not appear to have been more than sixty miles broad. Many instances of this kind might be cited, of local currents, of the southern flow of a stream along the coasts of China, and on into the Indian Ocean, while outside of the myriads of islands, the Japanese Black Stream is moving in majestic circles, and in a contrary direction.

In another part of this work, I have cited a case of the drift of pumice and ashes, easterly from Java to Ponape, flowing just between, and in a contrary direction to, the sweep of the two great ocean currents, the Black Stream of the North, and the Peruvian current of the South Pacific.

In regard to this floating pumice, a late authority, speaking of a certain formation found on the bed of the ocean, states, that everything seems to show that the formation of the clay is due to the decomposition of fragmentary volcanic products, whose presence can be detected over the whole floor of the ocean. * * * The universal distribution of pumice, over the floor of the ocean, is very remarkable, and would at first appear unaccountable; but when the fact, that pieces of pumice have been known to float in sea water for a period of over three years, before becoming sufficiently water-logged to sink, is taken into consideration, it will be readily understood, how fragments of this material may be transported, by winds and currents, to an enormous distance from their point of origin, before being deposited upon the bottom.

Among the islands of the South Sea, the channels, as between islands, are free and clear, and carry deep, navigable waters, with probably few sunken rocks to interfere with navigation. The currents flow through some of these channels, varying with the localities, at the rates of ten, fifteen, twenty-one and thirty miles a day. For this reason, it is deemed best to lay well off from shore, when not in a good harbor, of atoll lagoon, or bay. Many vessels have been lost in the sweep of these island currents, dragging their anchors (where anchorage can be had, as very often deep water makes up to almost the reef-line), and drifting in on the breakers, completely at the mercy of the waves. This often happens, too, in perfectly clear weather, when there is no wind to aid the luckless navigator in "clawing off shore." The main currents, spoken of, have considerable depth, while in others their movements may be termed surface, and sometimes greatly influenced by winds and storms. Others may be termed deep sea currents, whose flow traverses the depths below. These are just as variable as the surface movements. Any bulky article, like a keg, weighted to sink to the depth desired, and with sounding-line and buoy attached, may sometimes be seen, carrying the buoy against the wind and surface current, at the rate of two miles an hour.

It will be readily seen that the course and speed of surface currents can be traced with greater facility than those flowing deep down in the sea. Although the custom is not general, still in the cause of science it should be so, that in all sea voyages, buoys or bottles, with complete data of time and place, should be cast adrift at least once a week during the voyage.

The data contained in bottle or buoy should, of course, contain the request to note time and place when recovered from the ocean. If this were a general practice among our mariners, the little messengers would be looked for with special interest. The valuable practical data coming from this little source alone would add greatly in helping to perfect current charts of the different oceans.

In view of the varying ocean streams, more particularly among the islands of the South Sea, should development and commerce go hand in hand, the idea of using auxiliary steam-power on all vessels engaged in this particular trade, should meet with some encouragement from the mercantile world. A great deal of time lost in the calms and currents of these regions might be saved, as well as certain protection from storms and adverse currents. In regions where the atolls are, only those experienced in navigating among them, can judge of the value steam-power would have, if only applied for a few hours. The lagoons of the atolls are always safe harboring, but how to reach them with a sailing-vessel in a dead calm, through narrow entrances, and with storms and currents threatening, with the sea breaking over the coral reefs on either hand, is still a problem for the sailor. The same difficulty, if we leave out the sudden gale and currents, presents itself in getting out. Even if the auxiliary were not made a part of the vessel, still a steam-launch of considerable capacity could be carried, to be used only when required. This, I am sure, would obviate many of the difficulties sailing-vessels have to encounter, when trading among the Pacific Islands.

The influence of the tides, mainly caused by the attractive force of the moon and the centrifugal force exercised by the earth's revolutions, no doubt affect the ocean currents considerably. Their rise and fall, ranging in some places from but a few inches to seventy feet, raising and lowering the ocean level alternately, create a variable system of currents too well understood by navigators to require an elaborate explanation here. If we admit, for example, that while we have a high tide on the one side of the earth, caused by the moon's attraction, and that directly opposite on the other side of the globe there is a high tide, the effect of the centrifugal force of the earth's revolutions, with the consequent depression of the water levels between these points, we have a simple explanation of high and low tides. These points are continually shifting, moving around the earth's watery surface, as the influences causing them move, and explain in a general way, if we leave out local influences, the world's tidal system.

The influence of the heat emitted from the great fire-belt nearly outlining the western hemisphere may have had considerable influence on the ocean currents of the Pacific. At a much earlier period in the world's slow geological processes, when its shell was many miles thinner, it is obvious that the heat from subterranean fires would be more readily imparted to the water, causing a flow of the colder portions towards the points where it had been expanded or driven away by the heat (much as we see the movements of the mobile element when heating it in a vessel over a fire). The impetus given in this manner to the ocean's flow, it would seem but natural for the heat of the sun's rays together with the prevailing winds and tides, to keep it in constant motion and agitation. Many authorities attribute the movement of ocean currents altogether to the influence of the winds, whose great force and power will be better understood by consulting the following table, compiled from the latest observations of meteorologists:

VELOCITY AND FORCE OF WIND.

Miles per hour.	Feet per minute.	Pressure on a sq. ft. in lbs.	Description of wind.
1	88	.005	Barely observable.
2	176	.02	} Just perceptible.
3	264	.045	
4	352	.08	Light breeze.
5	440	.125	
6	528	.18	} Gentle, pleasant wind.
8	704	.32	
10	880	.5	Fresh breeze.
15	1,320	1.125	Brisk blow.
20	1,760	2.	Stiff breeze.
25	2,200	3,125	Very brisk.
30	2,640	4.5	} High wind.
35	3,080	6.125	
40	3,520	8.	Very high wind.
45	3,960	10.125	Gale.
50	4,400	12.5	Storm.
60	5,280	18.	Great storm.
80	7,040	32.	Hurricane.
100	8,800	50.	Tornado.

The cause of prevailing winds, blowing over the surface of the Pacific and other oceans—the "trades"—is given by the best authorities, as naturally following from the differences occurring between what is termed the regions of high and low barometer. This, in plainer terms, and those that may be understood by children, (if my little work should be honored by their perusal), would be the difference—in the pressure or weight of the atmosphere. North and south of the equator lie the regions of the greatest pressure of the air, and in between these, there is a broad space, following the equatorial line, belting the world. This is the region of low barometer, or where generally the pressure of the air is lightest. Towards this belt, the air from the north and south flows, as naturally as a greater height or level of water would run towards that of a lower level. On the Pacific Ocean, these air currents come from northeast and southeast, curving and blowing west near the equator, forming the trade winds. Aristotle, who was probably the first to predict changes in the weather, much in the form of the meteorological predictions of the present day, must have had some idea of atmospheric pressure, and the differences occurring in its weight. A definite explanation, though, of the pressure of the air, was not had until 1643, when it was discovered and explained by Torricelli, a pupil of Galileo. The theory of trade winds was explained by George Hadley, about 1735. Humboldt's Treatise on Isothermal Lines was not published until 1817.

MONSOONS.

Maury says: Monsoons are, for the most part, trade winds deflected. When, at stated seasons of the year, a trade wind is turned out of its regular course, as from one quadrant to another, it is regarded as a monsoon. The African monsoons of the Atlantic, the monsoons of the Gulf of Mexico, and the Central American monsoons of the Pacific, are, for the most part, formed of the trade winds, which are turned back or deflected, to restore the equilibrium which the overheated plains of Africa, Utah, Texas and New Mexico have disturbed. Thus, with regard to the northeast and southwest monsoons of the Indian Ocean, for example—a force is exerted upon the northeast trade winds of that area, by the disturbance which the heat of summer creates in the atmosphere over the interior plains of Asia, which is more than sufficient to neutralize the forces which cause those winds to blow as trade winds; it arrests them, and turns them back; but, were it not for the peculiar condition of the lands about that ocean, what are now called the northeast monsoons, would blow the year round; there would be no southwest monsoons there, and the northeast winds, being perpetual, would become all the year what, in reality, for several months they are—viz., northeast trade winds.

EFFECT OF MONSOONS.

Upon India and its seas, the monsoon phenomena are developed on the grandest scale. They blow over all that expanse of northern water, that lies between Africa and the Phillippine Islands. Throughout this vast expanse, the winds that are known as the northeast trades, are here called northeast monsoons, because, instead of blowing from that quarter for twelve months, as in other seas, they blow only for six. During the remaining six months, they are turned back, as it were, for instead of blowing toward the equator, they blow away from it, and instead of northeast trades, we have southwest monsoons.

The monsoon is an innocent, peaceable breeze, and in no way related to the typhoon, that terror of navigators, in some parts of the Pacific. In fact, as Maury says, in his Sailing Directions, it is a curious thing, this influence of islands in the trade wind region, upon the winds of the Pacific. Every navigator who has cruised in those parts of that ocean, has often turned, with wonder and delight, to admire the gorgeous piles of cumuli, heaped up in the most delicate and exquisitely beautiful masses, that it is possible for fleecy matter to assume. Not only are these cloud-piles found capping the hills among the islands, but they are often seen to overhang the lowest isle of the tropics, and to even stand above coral patches and hidden reefs, "a cloud by day," to serve as a beacon to the lonely mariner out there at sea, and to warn him of shoals and dangers, which no lead or seaman's eye has ever seen or sounded out. These clouds, under favorable circumstances, may be seen gathering above the low coral island, and performing their office, in preparing for vegetation and fruitfulness, in a very striking manner. As they are condensed into showers, one fancies that they are a sponge of the most delicately elaborated material, and that he can see, as they "drop down their fatness," the invisible but bountiful hand aloft, that is pressing it out.

TYPHOONS.

Under this head, for brevity's sake, all those terrible phenomena, known as hurricanes, tornadoes and cyclones, generally applied to storms taking place over the land, might be included the disastrous gales of the Pacific, known as typhoons. True, the ocean was well named by Magellan, and no doubt exhibits less stormy proclivities than any of the mighty wastes of water, nearly covering the globe.

Maury admits the research and ability of Redfield in America, Reid in England, Tom of Mauritius and

Paddington of Calcutta, in explaining typhoons, stating the theory of this school: That these are rotary storms; that they revolve against the hands of a watch in the southern hemisphere; that nearer the center or vortex the more violent the storm, while the center itself is a calm, which travels sometimes a mile or two an hour, and sometimes forty or fifty; that in the center the barometer is low, rising as you approach the periphery of the whirl; that the diameter of these storms is sometimes a thousand miles, and sometimes not more than a few leagues; that they have their origin somewhere between the parallels of 10 deg. and 20 deg. north and south, traveling to the westward in either hemisphere, but increasing their distance from the equator until they reach the parallels of 25 deg. or 30 deg., when they turn toward the east, or "recurvate," but continue to increase their distance from the equator—that is, they first travel westwardly, inclining toward the nearest pole; they then recurve and travel eastwardly, still inclining toward the pole; that such is their path in both hemispheres, etc.

THEIR EXPLANATION.

Maury doubts the correctness of the above statements in many ways, yet does not prove anything to the contrary. Their inception, with the destructive forces they exercise *when fully under way*, seems to be derived exactly from the same natural laws that create the trade winds, the flow of the heavier air to occupy the space opened up by the lighter. The more rarified atmosphere may come from the influence of the sun's heat, the rapid evaporation of water from the ocean's surface, or from the rapid condensation of aqueous vapor and consequent fall of rain, which is always accompanied by rarification and liberation of heat, These sudden gyrating storms, of which the Atlantic, Pacific and Indian Oceans furnish many examples every year, have yet to be fully explained, as some of them are accompanied by all the phenomena of lightning, thunder and copious showers 'of rain. This may add to the many theories already advanced, the broad explanation that electrical changes and influences will supply. The many examples recorded of the destructive forces of these storms, these *myriads of whirlwinds*, traveling across sea or land, are too well known to need repetition here.

From Birt's Hand-Book of Storms, furnishing a record of hurricanes for a year's time, I find the West Indies credited with 113; South Indian Ocean, 53; Mauritius, 53; Bay of Bengal, 30; and the China Sea (Pacific), 46.

RAINFALL OF THE PACIFIC ISLANDS.

With but very few exceptions, the *island* world is bountifully supplied with rain. True, some are without sufficient moisture apparently, although the profuse vegetation throughout the different groups testify to an abundant supply. About the only exceptions are some of the volcanic rocks and guano islands, whose bare surfaces, have not the requisites for attracting moisture.

At the Aleutian Isles there is more than enough, while Queen Charlotte and Vancouver Islands abound in running streams from bounteous rains. The islands along the coasts of California and Mexico are not so

fortunate in this respect; but unimportant as they are
in size and products, this want is only felt locally, and
is not of general importance. The Hawaiian group,
with a rainfall of thirty-six inches per annum, may be
said to have an abundance. The Galapagos, and islands
further south, have barely a sufficiency, until the islands
of Southern Chili are met, and clear on to Cape Horn,
where the rainfall often reaches 200 inches in a year.
The islands of Juan Fernandez, Mas-a-Fuera, ·Pit-
cairn, the Paumatous, Society, Fiji, Friendly, Samoa,
Marquesas, New Caledonia, the Marshall system,
etc., attest, by the profusion of natural vegetation, an
abundant rainfall. New Zealand, Tasmania and Australia have copious showers, though the latter, with her
immense interior wastes, with their great evaporating
powers, leaves surface water scarce. The Solomon Archipelago, Santa Cruz, the New Guinea, Ireland, Britain,
Admiralty islands and groups, have abundant moisture.
Java, Celebes, Borneo, the Molluccas and Sumatra
are in some localities, too well supplied, the fall of
rain in parts of Sumatra and Borneo being from
100 to 200 inches per annum. Still further west,
and in the northeast part of the Bay of Bengal,
among the Khasi Hills, it is said that the mean record
of rainfall for twenty years is something like 493.19
inches per annum, claimed to be the greatest recorded
rainfall on the globe. The islands of the Chinese
Empire, as well as the Phillippines and Japan, are all
in the range of abundant precipitation. In fact,
throughout the islands of the Pacific, water has never
been a drawback. True, in some spots surface moisture is scarce, yet in nearly every case, where sinking
has been resorted to, a plentiful supply of fresh water
has been met with.

PORTS AND HARBORS.

The following exact geographical location of some of the principal harbors and ports of the Pacific islands, are taken from lists in the United States Hydrographic Office :

Anger, Java.—Fourth Point Lighthouse (time ball). 6 deg. 4 min. 18 sec., S. lat.; 105 deg. 53 min. 0 sec., E. long.   Netherlands Hydrographic Office.

Austral (Tubuai) Islands.—Rouroutou Island, North Point.   22 deg. 29 min. 0 sec., S. lat.; 151 deg. 23 min. 41 sec., W. long.   Kulczki.

Acapulco, Mexico.—Northwest angle of Fort.   16 deg. 50 min. 56 sec., N. lat.; 99 deg. 55 min. 28 sec., W. long.   Commmander Philip, U. S. N.

Australia, Sydney.—Observatory.   33 deg. 51 min. 41 sec., S. lat.; 151 deg. 12 min. 39 sec., E. long. Nautical Almanac.

Australia, Adelaide Port.—Snapper Point.   34 deg. 46 min. 50 sec., S. lat.; 138 deg. 31 min. 0 sec. E. long.   Australia Directory.

Australia, Melbourne.—Observatory.   37 deg. 49 min. 53 sec., S. lat.; 144 deg. 58 min. 42 sec., E. long.   Nautical Almanac.

Bandger Massin, Borneo.—Residency flag-staff. 3 deg. 18 min. 55 sec., S. lat.; 114 deg. 35 min. 8 sec., E. long.   Netherlands Hydrographic Office.

Batavia, Java.—Observatory (time ball).   6 deg. 7 min. 40 sec., S. lat.; 106 deg. 49 min. 7 sec., E. long.   Netherlands Hydrographic Office.

Barrow Point, Alaska.— Highest latitude of the United States.   71 deg. 27 min. 0 sec., N. lat.; 156 deg. 15 min. 0 sec., W. long.   Capt. Beechey, R. N.

Bonin Islands, Peel Island.—Port Lloyd Observatory. 27 deg. 5 min. 37 sec., N. lat.; 142 deg. 11 min. 30 sec., E. long. Commodore Rodgers, U. S. N.

Caroline Islands, Hogoleu.—North end of Isis Islet. 7 deg. 18 min. 30 sec., N. lat.; 151 deg. 56 min. 30 sec., E. long. Captain Simpson, R. N.

Christmas Island.—North Point of Cook Islet. 1 deg. 57 min. 17 sec., N. lat.; 157 deg. 27 min. 46 sec., W. long. Captain Skerrett, U. S. N.

Fanning Island. — Flag-staff, entrance to English Harbor. 3 deg. 51 min. 26 sec., N. lat.; 159 deg. 23 min. 35 sec., W. long. English survey.

Farallone Islets, California. — Lighthouse, South Islet. 37 deg. 41 min. 49 sec., N. lat.; 123 deg. 0 min. 4 sec., W. long. U. S. Coast and Geodetic Survey.

Fiji Islands.—Vanua Lavu Island, M'bua Bay, Dimba, Dimba Point. 16 deg. 48 min. 10 sec., S. lat.; 178 deg. 26 min. 14 sec., E. long. Findlay South Sea Directory.

Fiji Islands.—Viti Lavu Island, Summit of Malolo Islet. 17 deg. 44 min. 45 sec., S. lat.; 177 deg. 9 min. 0 sec., E. long. English survey.

Friendly Islands.—Tonga-Tabu Island, Nukalofa; King's Garden. 21 deg. 8 min. 20 sec., S. lat., 175 deg. 8 min. 7 sec., W. long. Lieut. Heath, R. N.

Formosa Island.—Kelung Harbor, south shore. 25 deg. 8 min. 25 sec., N. lat.; 121 deg. 45 min. 30 sec., E. long. Captain Collinson, R. N.

Galapagos Islands.—Charles Island, summit (1,780 feet). 1 deg. 19 min. 0 sec., S. lat.; 90 deg. 28 min. 0 sec., W. long. Captain Fitzroy, R. N.

Galapagos Islands.— Abingdon Island, summit (1,950 feet). o deg. 34 min. 25 sec., N. lat.; 90 deg. 44 min. 10 sec., W. long. Captain Fitzroy, R. N.

Gilbert or **Kingsmill Islands.**—Aurorai or Hurd's Island, South Point. 2 deg. 40 min. 54 sec., S. lat.; 177 deg. 1 min. 13 sec., E. long. Findlay, South Pacific.

Hainan Island (China).—Cape Bastian extreme. 18 deg. 9 min. 30 sec., N. lat.; 109 deg. 33 min. 30 sec., E. long. China Sea Directory.

Juan Fernandez Island.—Fort S. Juan Bautista. 33 deg. 37 min. 36 sec., S. lat.; 78 deg. 49 min. 45 sec., W. long. English survey.

Ladrone or **Mariana Islands.**—Ascension Island, Crater (2,600 feét). 19 deg. 45 min. o sec., N. lat.; 145 deg. 30 min. o sec., E. long. Captain Sanchez, Spanish Navy.

Louisade Archipelago.—St. Aignan Island, summit. 10 deg. 42 min. o sec., S. lat.; 152 deg. 43 min. o sec., E. long. Australia Directory.

Loyalty Islands.—Mare or Britania Island, South Point. 21 deg. 42 min. o sec., S. lat.; 168 deg. o min. o sec., E. long. Admiral D'Urville, French Navy.

Manila (Island of Luzon, Phillippine Group) **Cathedral.**—14 deg. 35 min. 31 sec., N. lat.; 120 deg. 58 min. 3 sec., E. long. Lieut. Commanders Green and Davis, U. S. N.

Marquesas Islands.—Nuka Hiva Island, Port Tai-o-hae, French Hill. 8 deg. 54 min. 11 sec., S. lat.; 140 deg. 5 min. 6 sec., W. long. Lieutenant J. E. Craig, U. S. N.

Mas-a-fuera Island.—Summit (4,000 feet).   33 deg. 46 min. 0 sec., S. lat.; 80 deg. 46 min. 0 sec. W. long.   H. M. S. *Albatross.*

Marshall Islands.—Arhuo Atoll, Northeast Point. 7 deg. 9 min. 17 sec., N. lat.; 176 deg. 56 min. 30 sec., E. long.   Commander Meade, U. S. N.

Mulgrave Islands.—Port Rhiu, north side of entrance.   6 deg. 14 min. 0 sec., N. lat.; 171 deg. 46 min. 0 sec., E. long.   Captain Berard, French Navy.

Mazatlan, Mexico.—Signal station. 23 deg. 11 min. 17 sec., N. lat.; 106 deg. 26 min. 39 sec., W. long. Commander Dewey, U. S. N.

New Britain.—Blanche Bay, Matupi Island, Northeast Point.   4 deg. 13 min. 20 sec., S. lat.; 152 deg. 10 min. 18 sec., E. long.   German survey.

New Caledonia.—Harbor of Noumea, Lighthouse at office of Captain of the Port.   22 deg. 16 min. 20 sec., S. lat.; 166 deg. 27 min. 8 sec., E. long. Lighthouse List.

New Guinea.—Cape Cretin, Cretin Islets.   6 deg. 43 min. 0 sec., S. lat.; 147 deg. 53 min. 20 sec., E. long.   Captain Moresby, R. N.

New Hebrides Islands.—Aniteum Island, Port Aniteum, Sand Islet.   20 deg. 15 min. 17 sec., S. lat.; 169 deg. 44 min. 44 sec., E. long.   Captain Denham, R. N.

New Hebrides Islands.—Tanna Island, Port Resolution, Mission.   19 deg. 31 min. 17 sec., S. lat.; 169 deg. 27 min. 30 sec., E. long.   Captain Denham, R. N.

New Ireland.—Carteret Harbor, Cocoanut Islet.   4 deg. 41 min. 26 sec., S. lat.; 152 deg. 42 min. 25 sec., E. long.   Captain Belcher, R. N.

North Islands.—Queen Charlotte Island, North Point. —54 deg. 20 min. 0 sec., N. lat.; 133 deg. 0 min. 0 sec., W. long. English.

New Zealand.—Queenstown, U. S. Transit-of-Venus Station. 45 deg. 2 min. 7 sec., S. lat.; 168 deg. 40 min. 6 sec., E. long. Captain Raymond, U. S. A.

Nagasaki, Jápan.—North angle of Custom-house Sea-wall. 32 deg. 44 min. 35 sec., N. lat.; 129 deg. 52 min. 9 sec., E. long. Lieut. Commanders Green and Davis, U. S. N.

Panama.—South Tower of Cathedral. 8 deg. 51 min. 12 sec., N. lat.; 79 deg. 32 min. 12 sec., W. long. Lieut. Commander Green, U. S. N.

Paumatou Islands (Low Archipelago).—Arnanu or Muller Island, Southwest Point. 17 deg. 53 min, 20 sec., S. lat.; 140 deg. 50 min. 26 sec., W. long. Commaisance des Temps.

Phœnix Islands.—Gardner's Island, center. 4 deg. 47 min. 42 sec., S. lat.; 174 deg. 40 min. 18 sec., W. long. Commander Wilkes, U. S. N.

Pitcairn Island.—Village. 25 deg. 3 min. 37 sec., S. lat.; 130 deg. 8 min. 37 sec., W. long. Captain Beechy, R. N.

Samoan Islands.—Savaii Island, Paluale Village. 13 deg. 45 min. 0 sec., S. lat.; 172 deg. 17 min. 0 sec., W. long Commander Wilkes, U. S. N.

Samoan Islands.—Upolu Island, Apia Harbor, Rugis Wharf. 13 deg. 48 min. 56 sec., S. lat.; 171 deg. 47 min. 34 sec., W. long. Captain Richards, R. N.

Santa Cruz Islands.—Vanikoro, Ocili Village. 11 deg. 39 min. 30 sec., S. lat.; 166 deg. 55 min. 10 sec., E. long. Admiral D'Urville.

Society Islands.—Boru-boru Island, Otea Vanua Village. 16 deg. 31 min. 35 sec., S. lat.; 151 deg. 46 min. 0 sec., W. long. Findlay.

Society Islands.—Tahiti Island, Papiete Harbor, Motu-uta Islet. 17 deg. 31 min. 39 sec., S. lat.; 149 deg. 34 min. 16 sec., W. long. Connaisance des Temps.

Solomon Islands.—Bougainville Island, Northeast Point. 5 deg. 30 min. 0 sec., S. lat.; 155 deg. 17 min. 14 sec., E. long. Admiral D'Urville.

Sumatra, Padang.—Apenberg flagstaff. 0 deg. 58 min. 1 sec., S. lat.; 100 deg. 20 min. 13 sec., E. long. Netherlands Hydrographic Office.

Singapore.—Fullerton Battery. 1 deg. 17 min. 11 sec., N. lat.; 103 deg. 51 min. 15 sec,, E. long. Lieut. Commanders Green and Davis, U. S. N.

Sandwich Islands.—Hawaii, Hilo Bay Lighthouse, 19 deg. 45 min. 0 sec., N. lat.; 155 deg. 5 min. 0 sec., W. long. Light House List.

Sandwich Islands.—Oahu, Honolulu, Reef Lighthouse. 21 deg. 17 min. 55 sec., N. lat.; 157 deg. 52 min. 13 sec., W. long. Hawaiian Government Survey.

Sitka, Alaska.—Middle of parade ground. 57 deg. 2 min. 52 sec., N. lat.; 135 deg. 19 min. 31 sec., W. long. U. S. Coast and Geodetic Survey.

San Francisco, Cal.—Presidio Station. 37 deg. 47 min. 30 sec., N. lat.; 122 deg. 27 min. 49 sec., W long. U. S. Coast and Geodetic Survey.

San Diego.—Coast Survey Astronomical Station. 32 deg. 43 min. 6 sec., N. lat.; 117 deg. 9 min. 40 sec., W. long. U. S. Coast and Geodetic Survey.

Tasmania.—Hobart Town, Transit of Venus Station. 42 deg. 53 min. 25 sec., S. lat..; 147 deg. 20 min. 7 sec., E. long. Professor Harkness, U. S. N.

Union (Tokalau) Islands.—Nuku-Nono, or Duke of Clarence Island, Southeast Point. 9 deg. 11 min. 15 sec., S. lat..; 171 deg. 37 min. 2 sec., W. long. Commander Wilkes, U. S. N.

Valparaiso, Chili.—Site of San Antonio Fort. 33 deg. 1 min. 53 sec., S. lat.; 71 deg. 38 min., W. long. English Survey.

Vancouver Island.—Esquimalt, Lighthouse. 48 deg. 25 min. 40 sec., N. lat. 123 deg. 27 min. 20 sec. W. long. Admiralty Light List.

Yokohama, Japan.—Flag-staff English naval storehouse. 35 deg. 26 min. 24 sec., N. lat.; 139 deg. 39 min. 14 sec., E. long. Lieut. Commanders Green and Davis, U. S. N.

CHAPTER XIX.

HISTORICAL AND BIOGRAPHICAL.

> And now, rejoicing in the prosperous gales,
> With beating heart, Ulysses spreads his sails ;
> Placed at the helm he sate, and marked the skies,
> Nor closed in sleep his ever-watchful eyes.
> There viewed the Pleiads and the Northern Team,
> And great Orion's more refulgent beam,
> To which, around the axle of the sky,
> The Bear, revolving, points his golden eye ;
> Who shines exalted on the ethereal plain,
> Nor bathes his blazing forehead in the main.
>
> POPE'S (Homer's Odyssey.)

THE first nation or people in the world, to make any practical progress in navigation and commerce, or carry on any considerable traffic, making the seas and oceans serve as a highway, were—with the possible exception of the Chinese—the Phœnicians. Our accounts of them date as far back as 2800 years before the Christian era. Phœnicia was one of the smallest countries of antiquity. It occupied that part of the Syrian coast, which stretches from Aradus (the modern Ronad) on the north, to a little below Tyre on the south—a distance of about fifty leagues. Its breadth was much less, being, for the most part, bounded by Mount Libanus to the east, and Mount Carmel on the south. The surface of this

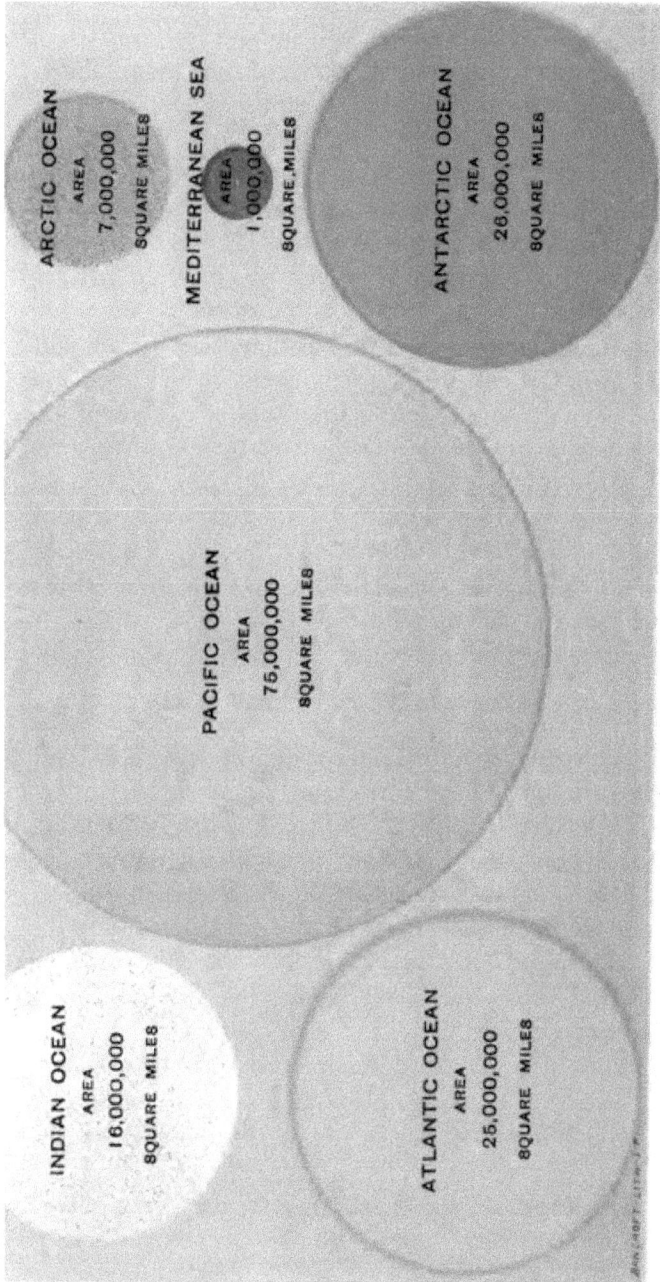

ARCTIC OCEAN
AREA
7,000,000
SQUARE MILES

MEDITERRANEAN SEA
AREA
1,000,000
SQUARE MILES

ANTARCTIC OCEAN
AREA
26,000,000
SQUARE MILES

PACIFIC OCEAN
AREA
75,000,000
SQUARE MILES

INDIAN OCEAN
AREA
16,000,000
SQUARE MILES

ATLANTIC OCEAN
AREA
25,000,000
SQUARE MILES

COMPARATIVE AREA OF THE WORLD'S GREATER BODIES OF WATER. (*estimated*)

OF THE PACIFIC OCEAN

narrow tract was generally rugged and mountainous, and the soil in the valley, though moderately fertile, did not afford sufficient supplies of food, to feed the population. Libanus and its dependent ridges were, however, covered with timber, suitable for ship building; and besides Tyre and Sidon, Phœnicia possessed the ports of Tripoli, Byblos, Berytus, etc. In this situation, occupying a country unable to supply them with sufficient quantities· of corn—hemmed in by mountains and powerful and warlike neighbors, on the one hand, and having, on the other, the wide expanse of the Mediterranean, studded with islands, and surrounded by fertile countries, to invite the enterprise of her citizens—they were naturally led to engage in maritime and commercial adventures, and became the boldest and most experienced mariners, and the greatest discoverers, of ancient times.

MERCHANTS OF THE OLDEN TIME.

From the remotest antiquity, a considerable trade seems to have been carried on, between the Eastern and Western worlds. The spices, drugs, precious stones, and other valuable products of Arabia and India, have always been highly esteemed in Europe, and have been exchanged for the gold and silver, the tin, wines, etc., of the latter. At the first dawn of authentic history, we find Phœnicia the principal centre of this commerce.

THE PHŒNICIANS.

Her inhabitants are designated, in the early sacred writings, by the name of Canaanites—a term which, in the language of the East, means merchants. The products of Arabia, India, Persia, etc., were originally conveyed to her by companies of traveling merchants, or caravans, which seem to have performed exactly the same part, in the commerce of the East, in the days of Jacob, that they do at present. (Genesis, xxxvii, 25, etc.) At a later period, however, in the reigns of David and Solomon, the Phœnicians, having formed an alliance with the Hebrews, acquired the ports of Elath and Eziongeber, at the northeast extremity of the Red Sea. Here they fitted out fleets, which traded with the ports on that, and probably with those of Southern Arabia, the west coast of India, and Ethiopia. The distance of the Red Sea from Tyre being very considerable, the conveyance of goods from one to the other, by land, must have been tedious and expensive. To lessen this inconvenience, the Tyrians, shortly after they got possession of Elath and Eziongeber, seized upon Rhinoculura, the port on the Mediterranean, nearest the Red Sea. The products of Arabia, India, and adjacent countries, being carried thither, were then put on board ships, and conveyed, by a brief and easy voyage, to Tyre. If we except the transit by Egypt (overland), this was the shortest and most direct, and for that reason, no doubt, the cheapest channel, by which the commerce between Southern Asia and Europe could then be conducted. But it is not believed, that the Phœnicians possessed any permanent footing on the Red Sea, after the death of Solomon. The want of it does not, however, seem to have sensibly affected their trade, and Tyre continued, till the foundation of Alexandria, to be the grand emporium for Eastern products, with which it was abundantly supplied, by caravans from Arabia, the bottom of the Persian Gulf, and from Babylon, by way of Palmyra.

COMMERCE.

The commerce of the Phœnicians with the countries bordering on the Mediterranean, was still more
extensive and valuable. At an early period, they
established settlements in Cyprus and Rhodes. The
former was a very valuable acquisition, from its proximity, the number of its ports, its fertility, and the variety of its vegetable and mineral productions. Having
passed, successively, into Greece. Italy and Sardinia,
they proceeded to explore the southern shores of
France and Spain, and the northern shores of Africa.
They afterwards adventured upon the Atlantic, and
were the first people, whose flag was displayed beyond
the pillars of Hercules.

INVENTIONS AND MANUFACTURES.

Nor were the Phœnicians celebrated only for their
wealth, and the extent of their commerce and navigation. Their fame, and their right to be classed amongst
those who have conferred the greatest benefits on
mankind, rest on a still more unassailable foundation.
Antiquity is unanimous in ascribing to them the invention and practice of all those arts, sciences and contri-
vances, that facilitate the prosecution of commercial
undertakings. They are held to be the inventors of
arithmetic, weights and measures, of money, of the art
of keeping accounts, and, in short, of everything that
belongs to the business of a counting-house. They
were also famous for the invention of ship building
and navigation; for the discovery of glass; for their
manufacture of fine linen and tapestry; for their skill
in architecture, and in the art of working metals and

ivory; and still more, for the incomparable splendor and beauty of their purple dye.

But the invention and dissemination of these highly useful arts, form but a part of what the people of Europe owe to the Phœnicians.

It is not possible to say in what degree the religion of the Greeks was borrowed from theirs, but that it was to a pretty large extent, seems abundantly certain. Hercules, under the name of Melcarthus was the tutelar deity of Tyre, and his expeditions along the shores of the Mediterranean, and to the straits connecting it with the ocean, seem to be merely a poetical representation of the progress of the Phœnician navigators, who introduced arts and civilization, and established the worship of Hercules wherever they went.

The Greeks were, however, indebted to the Phœnicians, not merely for the rudiments of civilization, but for the great instrument of its future progress—the gift of letters. No fact in ancient history is better established than that a knowledge of alphabetic writing was first carried to Greece by Phœnician adventurers, and it may be safely affirmed that this was the greatest boon any people ever received at the hands of another.

Before quitting this subject, we may briefly advert to the statement of Herodotus with respect to the circumnavigation of Africa by Phœnician sailors. The venerable father of history states that a fleet fitted out by Necho, King of Egypt, but manned and commanded by Phœnicians, took its departure from a port on the Red Sea, at an epoch which is believed to correspond with the year 604, before the Christian era; and that keeping always to the right, they doubled

the southern promontory of Africa, and returned after
a voyage of three years to Egypt, by the Pillars of
Hercules. (Herod. lib, 4, p. 42.) Herodotus further
mentions that they related that in sailing around Africa, they had the sun on the right hand, or to the
north—a circumstance which he frankly acknowledges
seemed incredible to him, but which, as every one is
now aware, must have been the case, if the voyage was
actually performed.

Many learned and able writers, and particularly
Gosselin (Reherches sur la Geographic Systematique
et Positive des Anciens, vol. 1, p. 204–217), have
treated this account as fabulous. But the objections
of Gosselin have been successfully answered in an
elaborate note by Larcher (Herodote, vol. 3, pp. 458–
464, ed. 1802); and Major Rennel has sufficiently
demonstrated the practicability of the voyage. (Geography of Herodotus, p. 682.)

Without entering upon this discussion, we may
observe that not one of those who question the au-
thenticity of the account given by Herodotus, presume to doubt that the Phœnicians braved the boisterous seas on the coasts of Spain, Gaul and Britain,
and that they had partially, at least, explored the
Indian Ocean. But the ships and seamen that did
this much, might, undoubtedly, under favorable circumstances, double the Cape of Good Hope. The relation
of Herodotus has, besides, such an appearance of good
faith, and the circumstance which he doubts, of the
navigators having seen the sun on the right, affords
so strong a confirmation of its truth, that there really
seems no reasonable ground for doubting that the
Phœnicians preceded by 2,000 years Vasco de Gama
in his perilous enterprise.

(McCulloch, Dict. Com. and Commer. Nav., vol. 2.)

THE COMPASS.

It has not been my purpose to trace navigation from its early dawn to the present time, in more than a general way, as thorough research and investigation into this interesting subject would require a separate volume. History of both ancient and modern times is so replete with the commercial ventures of countries, with accounts of voyages, discoveries and traffic, both by sea and land, that little doubt is left of the early knowledge and use of the compass, the invention and perfection of which is generally ascribed to a more modern period.

The knowledge of the cardinal points, as well as the use, probably in a rude way. of that important little instrument. the compass, we can trace back among the Chinese for nearly 3,000 years B. C. The knowledge of the true north, with its curious attraction and influence on the magnetic needle, turning and holding it at nearly right angles with the points of the rising and setting of the sun, was not only familiar and commented upon, but was put in practical use by the ancients of the higher order of intelligence. like the Chinese. Phœnicians, Egyptians, Grecians and Japanese.

The almost exact position retained in the heavens by the North Star, with the universe apparently sweeping in vast circles around it, must at first have been used as a guiding-point, while it would have been natural to take the opposite point for the other course. The other two points, east and west, were no doubt taken from the rising and setting of the sun, thus giving in a perfectly natural way the cardinal points of the compass.

PROPERTIES OF THE MAGNETIC NEEDLE.

It has been supposed that many of the ventures performed by the ancients at sea were only made during the day, the mariners anchoring during the night, never venturing far from land, or a depth of water too great for anchorage. This theory, I do not think, will bear very close inspection, as stormy weather and a lee shore would have rendered any great voyages impossible. The knowledge and practice of the ancients in mining and working the metals must have been considerable, as many of the writings of the fathers of literature will testify. The properties of magnetic iron ore, the load-stone of ancient and modern times, its faculty of not only attracting iron and steel, but of imparting its polar peculiarities to these metals, must have formed a long and curious study, and ages may have passed before some genius first tried and tested, or discovered its unvarying tendency, when so placed as to be little retarded in its movements, of turning and placing itself at right angles with the rising and setting of the sun, and pointing to the north. These first experiments must have been made after the ore had been brought to a metallic form and the metal shaped in the form of a needle, much in use in the olden time for their knitting, embroideries, tapestries and lace-work, for which the ancients of the higher order were so justly celebrated.

BY WHOM INVENTED.

Yet it is the common opinion, in our modern day, that the compass and its uses was the invention of Flavio Gioja, a citizen of the once famous republic of

Amalphi, very near the beginning of 1300 A. D. Many who wrote long years previous to this period, give abundant evidences of its knowledge and uses. Thus the great Spanish antiquary, Antonio de Capomany, and the famous Raymond Lully, in writings published as early as 1272 A. D., go to show the exact uses made of the compass in navigation. In one place Lully says: "as the needle when touched by the magnet naturally turns to the north;" and again, in another portion of his writings, he says: "As the nautical needle direct mariners in their navigation;" leaving us with the impression, as they were writing of periods many years anterior to 1200 A. D., that the little compass was in common use among mariners and "those who go down to the sea in ships."

In addition to the evident theoretical knowledge, of a portion at least, of the world's geography had by the Egyptian Ptolemies, they possessed maps and charts of all the regions known at that time.

The voyages performed by Hanno, Hippeas and Pythias, many years previous to the Christian era, were not accomplished without considerable knowledge of geography and navigation.

There is distinct mention made, in Chinese history, of the compass points, not only at the time mentioned above (2634 B. C.), but on down to 121 A. D., and again in 265 and 419 A. D. The best authorities state, that the compass was *introduced* into Europe in 1184 A. D., while some writers ascribe its discovery to Gioja, at the commencement of 1300 A. D. Dr. Gilbert states, that it was introduced into Italy, by Marco Polo, in 1295. There is also evidence of its use in France in 1150, in Syria about the same time, and in Norway previous to 1266.

HOMER'S KNOWLEDGE OF SHIP BUILDING.

Now toils the hero; trees on trees o'erthrown,
Fall crackling around him, and the forests groan.
Sudden, full twenty on the plain are strow'd,
And lopp'd and lighten'd of their branchy load.
At equal angles these disposed to join,
He smoothed and squared them by rule and line.
(The wimbles for the work, Calypso found,)
With these he pierced them, and with clinchers bound.
Long and capacious, as a shipwright forms
Some bark's broad bottom, to outride the storms,
So large he built the raft; then ribb'd it strong,
From space to space, and nail'd the planks along;
These formed the sides; the deck he fashioned last;
Then o'er the vessel raised the taper mast,
With crossing sail-yards dancing in the wind;
And to the helm, the guiding rudder joined;
With yielding osiers fenced, to break the force
Of surging waves, and steer the steady course.
Thy loom, Calypso, for the future sails
Supplied the cloth, capacious of the gales.
With stays and cordage, last he rigged the ship,
And, roll'd on levers, launch'd her in the deep.

POPE'S (*Homer's Odyssey.*)

In the quotations presented, from the "Odyssey" of Homer—who was writing at a time something over eight hundred years previous to the Christian era, and describing events that took place about 1200 B. C.—a familiarity, not only with ship building, but an astronomical knowledge, and its uses in navigation, is displayed, that may justly excite wonder and admiration. Of the 1,152 ships Homer describes in the Iliad, as carrying troops, and participating in the Trojan war, not one is mentioned as relying solely on oars as a propelling power. All are described as sailing vessels, and under the guidance of experienced sailors and navigators, whose knowledge of navigation descended from previous ages

ANSON, LORD GEORGE.—Born in England 1697; died 1762. Was in command and served on the east coast of America; in 1739 was recalled on the outbreak of the Spanish war; in 1740, sailed from England with eight men of war, to harrass the Spaniards in the South Seas, by way of Cape Horn; crossed the Pacific with only one of his fleet, the *Centurion*, having lost most of his men through scurvy; made some valuable captures and discoveries among the Pacific Islands, in addition to contributing through his journals, surveys and charts, a great deal of information in regard to the Pacific; served successfully against the French in 1747, and was promoted through all the admiralty grades of the English Navy; was the author of a book very celebrated in its day, entitled "Lord Anson's Voyage Round the World."

BALBOA, VASCO NUÑEZ DE.—Spanish soldier and navigator. Born in Lapan (by some authorities at Xeres, de les Cavalleros, Estramadura, Spain) in 1475, and beheaded at Castilla de Oro Darien (or at Acla, near there), in 1517. He first sailed on the Atlantic with Bastidas, and afterwards with Enciso, agent of Ojeda, finally reaching Darien about 1513. Jealousy and dissension among the leaders resulted in leaving Balboa in command, with the return of his rivals to Spain, where misrepresentation caused an order to be issued for his recall to that country. Meantime, Balboa had made many friends, among whom were native chiefs of the isthmus. His love of adventure, with a desire to conciliate the Spanish king, urged him to new exploits and adventures. This resulted from information communicated by one of the native chiefs in his overland journey across the isthmus, and the discovery (to him) of the Pacific Ocean, September 26th, 1513 (September 25th, 1513—*Bancroft*), taking possession in the name of Spain. This resulted in his re-establishment in favor at court, and his creation to the rank of admiral and deputy governor. The viceroy Davila, of the new province, arrived at Darien some time in 1514. Jealousies and dissension between the commanders continuing, Balboa (whose energy and restless daring ill-fitted him for a life of political intrigue), with great enterprise and labor transported the timbers and other materials of his ships across the isthmus. Rebuilding his vessel on the Pacific shore (in 1515-16-17), sailing on the great sea and making many valuable discoveries, among others the Pearl Islands (and through tradition only), the wealth and location of Peru. Through the wiles of Davila, or Pedrarias, he was induced to return to Darien, and was beheaded, as a dangerous political offender. As Balboa is often credited with the discovery of the Pacific Ocean, it would be well to note (and not, however, with all due respect and admiration for the adventurous Spaniard) the voyages of the celebrated Venetian traveler, Marco Polo, in the 13th century, and the voyages and discoveries of Post Commander Antonio d'Albreu and Francisco Serram, who first saw and noted the island of Papua, or New Guinea, in 1511. The greatness of the man is too well established by history to require any additional glory from discoveries not justly belonging to him. The feat he performed in transporting the different parts of his vessels across a country, that, even to-day, is a labyrinth of foliage and a net-work of almost impassable

mountains, ravines, and swamps, has never been surpassed. Speaking of him, Herrera, who in his writings
is anything but enthusiastic, says: "No living man in
all the Indies dared attempt such an enterprise, or
would have succeeded in it, save Vasco Nuñez de
Balboa."

BEECHY, FREDERICK WILLIAM.—Born in London
in 1796; died in 1856. An English naval officer of
great ability. Served in English Navy during wars
with France and America. In 1818 he was with
Franklin, in Bucham's Arctic Expedition, and afterwards with Parry, in the voyage of the *Hecla;* served
several years in the Pacific Ocean, making many valuable surveys and discoveries; passed through Behring's Straits, reaching nearly 72 deg. north latitude.
A man of great practical attainments, he made many
valuable additions to geography, navigation, meteorology, hydrography, as well as some valuable contributions to literature; made rear-admiral in English
Navy in 1854, and President of the Geographical Society in 1855.

BANKS, SIR JOSEPH.—Born in London in 1743, and
died in 1820. Was a man of vast scientific attainments, explorer and voyager, from Labrador to New
Foundland, and from Iceland and the Hebrides, to the
Society Islands in the South Sea. He accompanied
Captain Cook in his first voyage to the Pacific, to observe the transit of Venus; his valuable services in
this voyage, occupying three years, opened up much
that was new and useful to the scientific world. His
discoveries, in natural history and botany, together
with many valuable drawings and specimens and vast

collection of books, he bequeathed to the British Museum He was made baronet in 1781, and received the Order of the Bath in 1795.

BOUGAINVILLE, LOUIS ANTOINE DE.—Born in Paris in 1729; died there in 1814. A celebrated author, politician, soldier and sailor, and the first French circumnavigator of the globe. Was with Montcalm in Canada, as aid-de-camp; set sail around the world in 1766, passing through the Straits of Magellan, and through the Paumatou group, discovering new islands, arriving at Tahiti April 6th, 1768; visited the Samoan group, naming them the Navigators, called at the New Hebrides, and made a partial survey of the east coast of Australia; sailed through the Louisades and along the Solomon Archipelago, and harbored at Port Praslin, New Ireland. From there, after repairing his ships, he skirted the northern coast of New Guinea, discovered some new islands, and through the Molluccas, the Indian Ocean, rounded the Cape of Good Hope, reaching St. Malo in 1769, after an absence of about two years and four months; published a 2-volume account of his voyage in 1771-2. In 1778, was in command in the French navy, and served against England, in the American War of Independence, with distinguished courage and ability. Planned several voyages to the Arctic Seas, but meeting with but little encouragement, resigned from the navy in 1790; was afterwards ennobled by Napoleon I.

BEHRING, VITUS—Born in Denmark in 1680, and died in 1741. He entered the Russian naval service in 1704, and was made captain by Peter the Great, for distinguished services. He was placed in command in 1725, of a voyage of discovery to the Arctic Seas; discovered the straits that bear his name, and the separation between Asia and America (in second voyage of 1728), outlining and surveying the coast of Siberia. He made a third voyage in 1741, on a North Polar expedition, reaching about 69 deg. north latitude, but owing to stress of weather and sickness among his crews, was compelled to return; was wrecked on Behring Island, in 55 deg. 22 min. north latitude, 166 deg. east longitude, where he died, after going through all the hardships that could befall a castaway in the desolate Polar Seas.

BYRON, JOHN.—Born November 8th, 1723, and died April 10th, 1786. Served with Anson as midshipman; was wrecked off the Patagonian coast, and lived on a desolate island in that region for five years (1740–46); publishing a narative of his sufferings in 1768; was placed in command of an exploring expedition in 1764, making some important discoveries. As an accomplished sailor, he had few superiors, and as an author, met with success. His sons also were men of mark and ability, culminating in his grandson, Lord Byron, the poet.

CARTERET, PHILIP.—Was captain of the *Swallow*, one of the vessels under Samuel Wallis, which sailed from England on a voyage of discovery to the South Seas, August 22d, 1766; his second voyage was on private account, discovering and naming Gower and Carteret Isles, Queen Charlotte Isles, Pitcairn, etc., rediscovering and naming the Admiralty group, and returning to England in 1769.

ENTRANCE TO MANILA HARBOR.

COOK, CAPTAIN JAMES.—Born in Yorkshire, England, October 27th, 1728, and killed at Owyhee (now Hawaii), one of the Sandwich Islands, February 14th, 1799. First served at sea in merchant line, entering the royal navy in 1755; was promoted rapidly through all the lower grades, and placed in command of the frigate *Mercury*, one of the squadron, co-operating with General Wolfe at Quebec. His services there, as navigator, pilot and soldier, were rewarded by a command of the flag-ship *Northumberland*. His surveys of the coast of Newfoundland and Labrador, with frequent publications of maps and charts, together with a minute account of his observations of an eclipse of the sun, placed him in the front rank, as a man of high attainments. In 1768, he sailed in command of the *Endeavor*, to observe the transit of Venus, from a position in the South Sea, selecting Tahiti, of the Society group, where he arrived April 13th, 1769. After successfully accomplishing the main object of the voyage, he set sail on a general voyage of discovery, re-locating New Zealand, taking possession of the Australian coast, near Botany Bay, surveying and charting some thirteen hundred miles of coast line, and establishing Australia as an island, as well as its separation from Papua. After many adventures and escapes, he returned to England in June 11, 1771, having sailed around the globe. In July 13th, 1772, he again sailed in command of the *Resolution*, and *Adventure*, to "circumnavigate the whole globe, in high southern latitudes, making traverses, from time to time, into every part of the Pacific Ocean, which had not undergone previous investigation, and to use his best endeavors to resolve the much agitated question of a southern continent." In this voyage, he reached 71 deg.

10 min. south latitude, in 106 deg. 54 min. west longitude. After wintering at the Society Islands, Cook
made some valuable surveys of the Pacific, between
Easter Island and the New Hebrides, discovering and
naming New Caledonia, etc. He returned to England,
by the Cape of Good Hope, July 30th, 1775, being
absent something over three years. In 1776, he vol-
unteered to conduct an expedition to discover a northwest passage to Asia, which he proposed to attempt,
by way of Behring Strait. Before sailing north, he
spent some time in voyaging among the islands of the
Pacific, discovering (it was supposed) the Sandwich
Islands, in 1778. Sailing north, along the coast of
North America, determining the most westerly portion
of that country, and its distance from Asia, he reached
Icy Cape, August 17th, 1778, where his further passage
was barred by the ice. Returning to Sandwich Islands
to winter, with the view of renewing the expedition
when the weather permitted, he discovered the islands
of Hawaii and Maui, of the Sandwich group. Having
lost one of his small boats in one of the inlets of Hawaii, stolen by the natives, he landed, with a lieutenant
and nine men, to recapture it—or one of the chiefs, as
hostage for its return; a fight ensued, and Cook, with
several of his men, were killed, their bones being
recovered a week afterwards. That Cook, and the
men killed with him, were devoured by the natives, is
uncertain.

CAVENDISH, SIR THOMAS.—Born in Suffolk, England, in 1560; died at sea in 1592. His first voyage
was to Virginia, in 1586; his second, was with three
vessels, passing the Straits of Magellan in 1587,
spending some time in surveys of the coast of South

America; although the expedition was of piratical and buccaneering tendencies, in which line they made quite a success, capturing several valuable Spanish vessels and burning and sacking the towns of Acapulco, Payta, etc. Cavendish then sailed across the Pacific to the Ladrone Islands, through the Indian Archipelago and Strait of Java, around the Cape of Good Hope, reaching England September 9th, 1588, being the third, to circumnavigate the globe; was knighted by Queen Elizabeth, and started on another voyage in 1591, which he failed to carry out on account of sickness, mutinous crews, and finally his death, on the homeward passage.

DAMPIER, WILLIAM.—Born in England in 1652; date of death uncertain. Sailor, soldier, author, pilot and buccaneer. Crossed the Isthmus of Darien in 1679, with a party of pirates, capturing several towns, pillaging and laying them in ruins; captured several Spanish vessels also, with which they sailed along the South American coast, robbing and destroying many seaport towns. In 1684 he accompanied Captain John Cook on a piratical expedition, along the coast of Chili, Peru and Mexico; afterwards crossed the Pacific Ocean, cruising among the islands of the Indian Archipelago, arriving in England in 1691; published a book, his "Voyage Around the World." In 1699, sailed from England in command of sloop of war, on a voyage of discovery in the South Seas, exploring the western coast of Australia, the coast of New Guinea, New Britain, New Ireland and the Molluccas. On returning, was wrecked off the island of Ascension, reaching England in 1701. Followed the sea up to 1711. He published also "A Treatise on

Winds and Tides," and a vindication of his voyage to the South Sea, in the ship *St. George*, in 1707.

DANA, JAMES DWIGHT.—Born in Utica, N. Y.. February 12th, 1813. An American mineralogist and geologist, and author, of great ability. In December. 1836, was appointed mineralogist and geologist to the American Exploring Expedition to the Southern Atlantic and Pacific Oceans, under Commodore Wilkes, sailing in 1838, and returning in 1842. His researches into the island formations of the South Sea, the shells, the coral, the volcanic formations. etc., show erudition and patient research, with practical observing powers seldom surpassed. His works and contributions to science have been valuable and voluminous, being accepted authority in all parts of the civilized world.

DARWIN, CHARLES ROBERT.—Born in Shrewsbury, England, February 12th, 1809; sailed with Captain Fitzroy, in the *Beagle*, in his voyage around the world. as naturalist, in 1831, returning in 1836. During this voyage, Darwin examined the greater part of the South American coast; many of the Pacific islands ; New Zealand and Australia being visited and examined, as well as Mauritius, in the Indian Ocean. An account of the voyage was published in 1839, Darwin contributing materially to the scientific value of the work. His works on coral reefs, volcanic islands, geology, zoology, with many other contributions to the cause of science, were followed by his "Descent of Man," and "Selection in Relation to Sex," which have probably given him his greatest celebrity, or notoriety.

DRAKE, SIR FRANCIS.—Born in England, in 1545;
(by some authorities, in 1539;) and died at sea, near
Puerto Bello, December 27th, 1595. His first expedition of any moment, was with Sir John Hawkins, in
naval engagements, along the Atlantic seaboard, and
in the Gulf of Mexico, with the Spaniards. While in
Central America, like the greater and better man, Balboa, he saw the waters of the majestic Pacific, from one
of the mountain peaks of the isthmus, resolving to
make the mighty sea the scene of his future exploits.
Receiving a roving commission from Elizabeth, in
1577, he sailed through the Magellan Straits, pillaging a portion of the coasts of Chile and Peru; sailing
for North America, arriving at California, at Drake's
Bay (now known to be a point, somewhat different
from the Bay of San Francisco), where he took possession of California, in the name of Queen Elizabeth,
in 1577. Having made some valuable captures from
the Spaniards, and fearing to return as he came, he
attempted the northeast passage to the Atlantic, but
was driven back by the cold weather and impassable
fields of ice. Sailing south, by Japan, the Phillippines,
and through the Mollucca Islands, and across the
Indian Ocean, he rounded the African cape, reaching
England on the 3d of November, 1580—the first Englishman to circumnavigate the globe. His success
met with speedy recognition by the Queen; leading,
finally, after many naval adventures on the Atlantic, to
his appointment as Vice-Admiral, under Lord Howard.
It has been supposed, that Drake was the discoverer
of California, as well as the Bay of San Francisco.
Where he landed, was Point Reyes—latitude, 37 deg.
59 min. 5 sec. north. Cabrillo is also credited with
the discovery, about 1542; he locating and naming

Cape Mendosa (now Mendocino). Cortez, in 1536, discovered the peninsula and Gulf of California.

CORTEZ (or CORTES), HERNAN (or HERNANDO).— Born in Medellin, Estramadura, in 1485, and died near Seville, December 2d, 1554. His first voyage of any note, was to San Domingo, and from there, in 1511, with Velasquez, to Cuba. He was appointed by the Governor to command an expedition to Mexico, to conquer and settle that country, which Grijalva, its Spanish discoverer, had failed to do. Cortez sailed from St. Iago in 1518, and landed on the coast in 1519. Founding the town of Vera Cruz, he burned his ships, and marched for the interior; after many hardships and reverses, he completely subdued and conquered Mexico, in the decisive battles, with the natives, of 1520–21. History and biography are so replete with this conquest, as well as of the minutest details of the life of the great Spanish adventurer, that but a short notice seems all that is necessary here. His discoveries on the Pacific, the Gulf of California, and its survey, the location of the Peninsula of California, together with several expeditions sailing under his patronage, entitle him to a place among the early navigators of the South Sea. His varying fortunes left him to die, as above, in solitude and despair.

FERNANDEZ, JUAN.—Navigator, pilot, and discoverer in the Pacific. In 1563, he first sighted the island now bearing his name (celebrated in the annals of "Crusoe"), and Mas-a-fuera, afterwards granted to him by the Spanish government. In 1574, he discovered the islands of San Felix and San Ambrose, making many voyages and discoveries in the South Sea, particularly the flow of the currents along the coast of South America. He is credited, sometimes, with being one of the early discoverers of New Zealand and Australia.

FITZROY, ROBERT.—Born in England, July 5th, 1805; died there, April 30th, 1865. Entered the navy in 1819, serving in the Mediterranean, and at South American stations; in 1831, was placed in command of the *Beagle*, making a voyage around the world, being accompanied by the celebrated Darwin, as naturalist and geologist of the expedition; in 1843, was appointed governor and commander of the colony of New Zealand, where he served for three years. He was the author of several works, contributing largely to meteorology, and establishing a system of storm warnings in 1862.

FRANKLIN, SIR JOHN.—Born in Lincolnshire, England, April 16th, 1786, dying in the Arctic regions, on June 11th, 1847 (as per records discovered by McClintock, in his expedition to the Arctic, in 1859). Franklin served in the English navy, as a midshipman, in 1801, and in 1802 accompanied Captain Flinders in a voyage to the South Sea, to survey the coasts of Australia, occupying two years for its accomplishment. On the return, they were wrecked off the coast of Australia, barely escaping with their lives, fifty days being spent on a barren, sandy islet, before relief arrived. On his return to England, served with Nelson at Trafalgar, as signal midshipman to the fleet, in 1805, and afterwards on the American coast in 1812–15. His first Arctic expedition was in 1818, in command of the *Trent* with Captain Buchan, of the *Dorothea*.

They reached as high as 80 deg. north latitude, but, on account of an accident to Buchan's ship, were forced to return. In 1819, was in command of an Arctic expedition overland from Hudson Bay, and in 1825 was in command of a similar expedition, which was carried through with marked ability. Was knighted in 1829, receiving honors from many parts of the world. Served in command on the Mediterranean, and in 1836–43, was made Governor of Tasmania, or Van Diemen's Land. In 1845, was placed in command of an Arctic expedition, to discover the northwest passage, being his fourth visit to that region. He was last seen by an American ship captain on July 26th, 1845, and his fate remained unknown up to McClintock's discovery, as above (1859).

HUMBOLDT, FREDERICK H. ALEXANDER, BARON VON. —Born in Berlin September 14th, 1769; died May 6th, 1859. One of the most celebrated men of his day. In the arts and sciences he was far advanced; a great leader in astronomy, finance, chemistry, natural philosophy, mineralogy, natural history and geography. He was in addition, one of the world's greatest travelers, making many journeys, overland and by sea. In Europe, including many thousands of miles of overland journeyings in Russia, in North and and South America, on both coasts; now in Brazil, again in Chili and Peru, surveying and marking out the sources of the Orinoco and Amazon rivers, or climbing the great peaks of Chimborazo and Pichincha. Always energetic, indomitable and untiring, his many intellectual attainments opened nature's secrets to him, which he read, as from a great book. Again in Mexico, and then in the United States, establishing the accepted theory of the great volcanic fire-belt, marking out the earth's surface in isothermal lines, so as to compare the world's varying climatic conditions, nothing escaped his wonderful observing powers, or was misapplied or mislaid in the vast storehouse of his wonderful memory. Crowning his life with that great work "Kosmos," he died full of days, honored and regretted all over the world.

MAGELLAN, FERNANDO.—Born at Oporto, Portugal in 1470, and killed at Mactan, a small island in the Phillippine group, April 27th, 1521. He made several voyages from Portugal to India, and the islands of the Eastern Archipelago. On August 10th, 1519, an expedition from Spain sailed under command of Magellan, to reach the Spice Islands by a western route. It is supposed by many authorities that Magellan, in the course of his maritime career, had met with an old map of South America, delineating a route across its southern portion, of which he availed himself in his voyage. His first attempt was by way of the Rio Plata. Failing in this, he skirted the shore until the ocean cut-off was reached and passed, in the latter part of 1520, naming Tierra del Fuego, and from the smooth waters first met with on the great ocean—the Pacific. Sailing north, he crossed the line on February 13th, 1521, reaching the Ladrones and Phillippines in March of that year. His great desire for the religious advancement of the natives at Mactan, where he insisted on baptising one of the chiefs and his followers, terminating in a quarrel, resulted in the death of Magellan The remainder of the expedition, under Caraballo, sailed for the Spice Islands, touching at Borneo and other islands of the Archipelago; finally, making a station at one of the Molluccas. Here, one of the vessels, *Victoria*, was put in repair, provisioned and placed under command of Sebastian del Cano (Magellan's pilot), who continued the voyage, reaching Spain in 1522, after an absence of nearly three years. This is the first circumnavigation of the globe of which there is authentic record.

KOTZEBUE, OTTO VON.—Born in 1787, and died in 1846. First sailed with the Russian Admiral Krusenstern around the world. He made his second voyage in command in 1815, for explorations in the Arctic and Pacific Oceans. Many islands were visited, and some discoveries made, returning in 1818. In 1823 he again sailed in command, visiting many of the more important island groups in the Pacific and the Russian settlements in Kamptchatka, and returning to Cronstadt in 1826. This latter voyage was one of vast importance, many corrections being made in the latitudes and longitudes of places, as well as additions to the botanical knowledge of the world, with much that threw light on the history of people of the countries visited.

KRUSENSTERN, ADAM J.—Russian navigator and admiral. Born in Esthonia, November, 1770, and died at Revel, August 24th, 1846. Served as midshipman in the war with Sweden, and afterwards with the English fleet, visiting America, China and India; sailed in command for Russia in 1803, with a view to extend and create commerce with the Asiatic countries, particularly China and Japan, returning in 1806; he voyaged by way of Cape Horn, and returned by the Cape of Good Hope, this being the first Russian expedition to sail around the world.

PERRY, MATTHEW CALBRAITH.—Born in Kingston, R. I., in 1795, and died in New York in 1858. Served in U. S. Navy as midshipman as early as 1809; was under Commodores Rodgers and Decatur; was made captain in 1837, and in command of the squadron on the coast of Africa, and of the fleet in the Gulf of Mexico, during the Mexican war. In 1852 he sailed in command of the expedition to Japan, where he distinguished himself in accomplishing an important treaty with that country in 1854.

PIZARRO, FRANCISCO.—Born in Spain in 1471, and was killed in a quarrel at Lima, Peru, June 26th, 1541. Was conqueror of Peru and the founder of Lima; served with Ojeda, Balboa, and afterwards under Pedrarias, governor of Darien; he made several expeditions along the coast of South America, but with no important results, except the knowledge he gained of the wealth and fertility of Peru. It was not until 1531, under commission from Charles V. of Spain, when he sailed for Peru, that he finally succeeded in the conquest of that country. He had considerable ability as a soldier, and was skillful as a navigator, although his voyages and discoveries were few and unimportant. His first knowledge of Peru, with the conquest of the land of the Incas, are elaborately detailed, in history and biography.

LA PEROUSE, JEAN F. DE GALAUP.—A French navigator; born August 22d, 1741, in France; died (supposed) at the island of Vanikoro, one of the Santa Cruz group, South Pacific, in 1788 or 1789. Entered the French navy at an early age, serving with varying fortunes against the English, and subsequently in the

American War of Independence. Under Louis VI., he fitted out the two frigates, *Astrolabe* and *Boussole*, and sailed for the Pacific August 1st, 1785, by way of Cape Horn. He explored the North American coast, from Mount St. Elias, Alaska, as far south as Monterey, sailing thence for Asia. In 1787 he partially surveyed the channels among the Phillippines, the China Sea, Japan, to the Russian possessions in the north, sending his charts, journals and observations to France. In the latter part of the same year, he sailed for the South Sea, touching at Maouna, one of the Samoa Islands, losing the commander of the *Astrolabe* and many men in a conflict with the natives. From here he sailed for Botany Bay, Australia, where he forwarded an account of his voyages and discoveries to the French minister, also explaining and mapping out his intentions for the future— dated at Botany Bay, February 7th, 1788. This was the last communication ever received from the French admiral, his fate remaining a mystery to-day. In 1791 a French squadron, under Admiral D'Entrecasteaux, sailed in search of the missing navigator, but failed in making any discoveries. D'Urville, who was at Hobart Town in 1828, learned through information brought by an American ship captain, of the remains of wrecks existing at Vanikoro Island. His researches brought to light the fact that Perouse's vessels had been wrecked on the reefs, and those of his crew who had not been drowned or murdered by the inhabitants, succeeded in building a small vessel from the wreck, and sailed for parts unknown. The anchors and cannons found at Vanikoro, afterwards taken to France, fully attest the unfortunate ending of the noble admiral's voyage.

POLO, MARCO.—A Venetian, sailor, author and traveler; born about 1254, and died in 1324. In 1271 Polo started with his father (Nicolo Polo, and Maffeo his uncle) on an overland tour of China and other countries of Asia, where many years of his life was spent in mercantile and other pursuits, as well as frequent journeyings throughout the Asiatic world located south of Russia. Before terminating his travels, one of the great desires of a busy life, was to continue his explorations by sea, which was gratified by his voyage to Japan, called in his day *Zipangu*. His return to his native land was by way of the Phillippine, Spice and the islands of Java, Sumatra, Borneo, Ceylon, Madagascar, etc., and some points on the east coast of Africa, thence back by way of the Arabian Sea, the Persian Gulf, and landing in Persia. The journey was continued overland to the Black Sea, where a vessel was obtained to convey them to Venice, arriving there in 1295. Polo's account of his voyages and experiences were received with general derision and doubt, although the full particulars of his adventures were not finished and published till 1298. His work was dictated to a fellow prisoner in Genoa, where Marco was held in durance for several years by the Genoese, as a prisoner of war, having served with the Venetians in an expedition against that country. Polo's work was regarded as a well concocted fable, and the slow processes of time, with gradual discoveries being made by sea and land, were necessary to prove the truth of his statements. His accounts of Cathay, Zipangu, the islands, the spices, silks and precious stones, met with in his wanderings, were received with grave deprecation and doubt, which the practical evidences of wealth brought back

with him could not shake. True, many of the things
related by him were painted with collateral writing
that came from tradition, much as we see in many
publications of the present time. Probably a better
idea of the meaning I wish to convey would be had
by an example of his writings (one evidently colored,
the other as near the truth as limited knowlege would
permit), taken from Murray's edition of "The travels
of Marco Polo," published in 1858. He (Polo) says:
Having described so many inland provinces, I will
now enter upon India, with the wonderful objects in
that region. The ships in which the merchants navigate thither are made of fir, with only one deck, but
many of them are divided beneath into sixty compartments, in each of which a person can be conveniently
accommodated. They have one rudder and four masts,
while some have two additional, which can be put up
and taken down at pleasure. Many of the largest
have besides as many as thirteen divisions in the hold,
formed of thick planks mortised into each other.
The object is to guard against accidents, which may
cause the vessel to spring a leak, such as striking on
a rock or being attacked by a whale. This last circumstance is not unusual, for during the night the
motion of the ship through the waves raises a foam
that invites the hungry animal, which, hoping to find
food, rushes violently against the hull, and often forces
in part of the bottom. The water entering by the leak
runs on to the well, which is always kept clear, and
the crew, on perceiving the occurrence, remove the
goods from the inundated division, and the boards are
so tight that it cannot pass to any other. They then
repair the injury and replace the articles. Again
in describing Japan and the myriads of islands of that

country, together with the Phillippine group and the Archipelago of Chusan, he says: You must know that the gulf containing this island (one of the large islands of Japan) is called that of Zin, meaning, in their language, the sea opposite to Manji. According to skillful and intelligent mariners who have made the voyage, it contains 7,448 isles, mostly inhabited. In all these there grows no tree which is not agreeably fragrant and also useful, being equal or superior in size to the lignum aloes. They produce also many and various spices, including pepper, white like snow, as well as the black. They yield also much gold, and various other wonderful and costly productions, but they are very distant and difficult to reach.* The mariners of Zai-tun and Kin-sai, who visit them, gain indeed great profits; but they spend a year on the voyage, going in winter and returning in summer, for the wind in these seasons blows from only two different quarters, one of which carries them thither and the other brings them back.† But this country is immensely distant from India. You may observe, too that though they be called Zin, it is really the ocean, just as we say the sea of England, the sea of Rochelle. The Great Khan has no power over

---

(* The number of islands stated is doubtless fanciful and exaggerated; yet when we consider the various groups composing the Oriental Archipelago, many consisting of numerous islets, the whole amount must be very great. They are, as we have justly noticed, productive beyond any other part of the world in aromatic and odoriferous plants, also very rich in gold.—*Murray, p. 243.*)

(† The distance would not be very formidable to a British mariner, but is otherwise to the ruder Chinese navigator; while this sea, too, is tempestuous and dangerous. The junks still perform only one voyage in the year, and as here correctly stated, sail in the winter, with the northeast monsoon, and return in summer with the southwest one.—*Marsden, p. 582.*)

these islands. Now let us return to Zai-tun, and re-sume our narrative. In truth, Polo so rounded up his narrative of travels and voyages, with the varied traditions and fireside tales of the countries he vis-ited (as his story of the "Griffin," the "Old Man of the Mountain," and "the birds carrying out the dia-monds, adhering to pieces of flesh thrown in the valley of Golconda"), that the many truths he re-lated seemed but a part of the fiction. Yet in geography, history, chronology, the manners and cus-toms of the people met with, he is in the main, correct. The grand results that may be said to have sprung from the travels and voyages of the wander-ing Venetian—the voyages of Columbus, Cabot, Ves-pucci, da Gama, and many other noted navigators—grew out of the writings of Polo. Of Columbus it is truly said, his aims were nothing less than the dis-covery of the marvelous province of Cipango, and the conversion to Christianity of the Grand Khan, to whom he received a royal letter of introduction. The main object of Columbus, his dreams and the-ories alike urging him on, was to reach the land of gold, spices, silks and precious stones, by a west-ern route, to find the land of Marco Polo—*Cipangu*. If we glance back fully *six hundred years*, and mark the course of the daring Venetian, or if we look around at the majestic grandeur of the New World, and credit him with but a portion of the results, the voyages of *later* discoverers, we should accord him the first page in the history of modern times.

QUIROS, PEDRO FERNANDO DE.—A celebrated Por-tuguese navigator, born in Elvere, Atlentejo, in 1560, and died in Panama, in 1614. . In 1595, joined the expedition of Alvaro Mendana, sailing from the New World, as navigator. Mendana had under him four vessels and four hundred men, it being his design to visit and colonize the Solomon Islands, discovered by him in a previous voyage, in 1567. In their voyage across the Pacific, he found, in addition to the discovery of some smaller islands, the group named by him Marquesas, in honor of the wife of Mendoza. Sailing from this cluster, they were caught in a tempest, damaging the fleet, and resulting in the loss of the admiral's vessel. Discouraged and disheartened at this misfortune, the crews mutinied, forcing Mendana to sacrifice many of the lives of his men and officers, and through remorse and regret, dying himself, September 17th, 1595. Quiros now took command of the expedition, and, after discovering many populous and fertile islands, proceeded to Manila, reaching that port, February 11th, 1596, with the squadron in a dilapidated and sinking condition. From Manila, Quiros returned to Mexico, and then to Peru, with a view to raising another expedition, to follow up the discoveries of Mendana and himself. Failing in this, he sailed for Spain, where his representations to Philip III, with the desire to discover the great Austral continent (Quiros probably being the first to represent its existence), resulted in his return to Peru, authorized to equip two vessels and a corvette. This being accomplished, he sailed from Callao, in command, with Louis de Vaes de Torres as second, December 21st, 1605. During the voyage, many islands were discovered, the Society group being among them, and getting a glimpse of what he supposed to be Australia, but afterwards proved to be the New Hebrides Islands. During a' violent storm, Quiros and Torres became separated; the former returning to

*21

Mexico in 1606, while the latter continued his voyage of discovery, to the north and west. In his voyage north, Torres discovered the straits that bear his name, and skirted the coast of New Guinea for eight hundred leagues.

Quiros still had a desire to discover and see the unknown land (Australia), and made another trip to Spain, to enlist royal favor in a new expedition. Failing in this, he returned to Panama, where his life passed away in futile efforts to accomplish dreams of new discoveries and conquests in the South Sea. The last of that coterie of daring soldiers and navigators of the sixteenth century, his life ebbed away within sight and sound of the surf waves of the Pacific. The memoirs of Quiros, addressed to Philip III, published in Seville in 1610, clearly depict the type of men, who gave Spain her former wealth and glory in the New World.

ROGERS, WOODS.—English navigator; in the Royal Navy in 1708, and sailed in command on a voyage around the world, from Cork Harbor, September 1st, 1708. After rounding Cape Horn, Rogers sighted, and made a landing at, the island of Juan Fernandez, January 31st, 1709. Captain Rogers relates: Our yawl, which we had sent ashore, did not return as soon as we expected; so we sent our pinnace (armed) to see the occasion of her stay. The pinnace returned immediately from the shore, and brought abundance of craw-fish, with a man clothed in goat-skins, who looked more wild than the first owners of them. He had been on the island four years and four months; his name was Alexander Selkirk, a Scotchman, who had been master of the *Cinque Ports Galley*, a ship

which came here with Captain Dampier, who told me
that this was the best man in her; so I immediately
agreed with him to be mate on board our ship. It was
he that made the fire last night, judging our ships to
be English.

Rogers continued his voyage from Juan Fernandez, by way of Guayaquil, the Galapagos, and the
North American coast, making several valuable captures of Spanish galleons. From California, the expedition sailed across the Pacific to the Phillippines, and
through the Molluccas, anchoring at Batavia. From
thence, across the Indian Ocean, around the Cape of
Good Hope, reaching the Thames, October 14th, 1711.
Captain Rogers wrote an account of his voyage around
the world, of which he says: This voyage being only
designed for cruising on the enemy, it is not reasonable to expect such accounts in it, as are to be met
with in books of travels relating to history, geography,
and the like. Something of that, however, I have in-
serted, to oblige the booksellers, who persuaded me
that this would make it more grateful to some sort of
readers. He died in 1732.

SAAVEDRA, ALVARO or ALONZO DE.—Was born
about the beginning of the sixteenth century. A relative of Hernando Cortez, whom he accompanied to
Mexico, he was alike a daring soldier and experienced
navigator. · Was sent in command of a small squadron, by Cortez, in 1526, for minor explorations in the
South Sea, and afterwards ordered by Spain to cross
the Pacific to the Spice Islands, on a voyage of discovery. Although the main object assigned, was the relief
of Garcia de Loaisa, who had sailed from Corunna, in
the track of Magellan, in 1525. He made some impor-

tant discoveries and observations during the voyage, adding much to the knowledge, slowly accumulating, in regard to the Pacific. Saavedra went down, with his vessel, in a hurricane, on the equator.

As something has already been said of Saavedra, in different parts of this work, briefly giving an insight into his daring character, it would be but a repetition to recall it here.

In 1529, Saavedra, returning towards New Spain, had sight of land in two degrees south, and ran along it above five hundred leagues, when he saw people of black, curled hair, called Papua; but, having sailed four or five degrees to the south, he returned toward the north, and discovered an isle, which he called the *Isle of Painted People*. And a little beyond it, in ten or twelve degrees, he found many low, small isles, full of palm trees and grass, which he called *los jardines*. The natives wear white cloth, made of grass; never saw fire; eat cocoas and fish, and dig boats with shells. Saavedra, perceiving the weather to be good, sailed towards the firm land and city of Panama, there to unload the cloves and merchandise he had, which might be carried in carts four leagues, to the River Chagres, which is said to be navigable into the North Sea, not far from *Nombre de Dios;* by which all goods might be brought a shorter way than round about the Cape of Good Hope.

(John Harris: Collection of Voyages and Travels. London, 1705. Page 272.)

SCHOUTEN, WILLIAM CORNELIUS.—A celebrated Dutch navigator, who died in 1625. He was the discoverer of the Schouten Islands, rediscovered by Carteret, who named them Admiralty. His principal voyage was in 1615, in command of the *Concordia*. An account of his expedition and adventures, in company with the intrepid Lemaire, was published in Amsterdam, in 1617.

SCHOUTEN, GAUTIER.—A Dutch navigator, who died in 1680. He was in the service of the Dutch East India Company, cruising principally among the islands of the East Indian Archipelago. A man of rare ability in his day, with a practical knowledge of the waters and islands of Western Oceanica, that served materially in establishing the foothold obtained by the Dutch in the Pacific. He published an account of his voyages, at Amsterdam, in 1676.

TASMAN, ABEL JANSEN.—Was born about 1600; time and manner of his death unknown. In the early part of his career, he served with the Dutch East India Company, in Japanese and Chinese waters, and later on, as a cruiser among the islands. In 1642, he was employed by the governor of the above company, to command in a voyage of discovery to the south of the line, and to ascertain the extent, if possible, of Australia, then known as New Holland. On the 24th of November, in the above year, he discovered the island of Tasmania, naming it Van Diemen's Land. The voyage, which occupied ten months, was one of some importance, as Tasman discovered New Zealand, the Fiji and Friendly groups, besides obtaining much valuable data in regard to Australia and New Guinea. He made a second voyage in 1644, with the intention of circumnavigating New Guinea and New Holland, of which there is no authentic data.

VANCOUVER, GEORGE.—Born in England in 1758 ; died there, May 10th, 1798. First sailed with Captain James Cook, in his second and third voyages ; was made lieutenant, and served for some years in the West Indies. His fourth voyage was made in command in 1791, to the British possessions in Western North America, which he reached, after touching at the Sandwich Islands, in 1792, when he took possession of Vancouver Island, and made many valuable charts from his surveys of the northern coasts, as well as the settlement of some complications that had sprung up, in regard to Vancouver Island. During his surveys of the northern coast, his winters were spent in the Sandwich Islands. Returned to England in 1795, surveyed and made many valuable notes of the west coast of South America, on his way back.

WILKES, CHARLES.—Born in New York in 1801 ; served as midshipman in the U. S. Navy, in the Mediterranean, in 1816, and on the Pacific in 1821–3. On August 18th, 1838, sailed in command of a United States exploring expedition, to the South Atlantic and Pacific Oceans, with five vessels and one store ship, visiting, surveying and exploring many islands of the Pacific, and with the many scientific men under him, making a valuable record, and important discoveries in both oceans. In 1840, the squadron arrived at the Fiji Islands and the Hawaiian group, where the scientific observations, maps and charts made, have contributed a great deal to a correct knowledge of the Pacific. In 1841, sailed to the northwest coast of America, partially exploring the Columbia and Sacramento Rivers, and the Bay of San Francisco. In the same year, sailed from the latter harbor, visiting the

Phillippines, Borneo, the Molluccas, Singapore, etc.,
crossing the Indian Ocean, rounding the Cape of Good
Hope, calling at the island of St. Helena, and other
points of interest in the Atlantic; reaching New York,
January 10th, 1842. Wilkes was the author of many
important works, while the voluminous records kept
of the expedition, and published by our Government,
contain an immense amount of valuable information.
Wilkes took part in the United States Civil War, serving with marked ability, and was created rear-admiral
on the retired list, July 25th, 1866.

CHAPTER XX.

ISLAND MISCELLANY, AND DEPTHS OF THE SEA.

Skill'd in the globe and sphere, he gravely stands,
And, with his compass, measures seas and lands.
DRYDEN (*Sixth Satire of Juvenal*).

THERE are many points of interest to be glanced
at, still, on the Pacific Ocean, a few of which I
note below, before concluding with the depths
of the sea.

NORFOLK ISLAND.

This island, located in latitude 28 deg. 58 min.
south, and longitude 167 deg. 46 min. east, something
over one thousand miles northeast from Sydney, has a
population, at present, of not over five hundred people, and an area of about fifteen square miles. It is
the principal of a group of small islets, known as the
Bird Islands. It is put down as one of Cook's discoveries, in 1774. The surface is extremely rugged,
standing high above the ocean level. In fact, so precipitous are its sides, that but two landing places are
to be found, indenting the shores, and these dangerous, from the baffling currents and heavy surf. A
portion of the lands, back from the coast, is very fertile, nearly all the products of tropical and temperate

ANCIENT RUINS—ISLAND OF ASCENSION.

regions growing luxuriantly. The island was not made a point of interest till 1787, when it was settled by convicts and ticket-of-leave men from Australia. In 1825, it was made a penal colony by that country, but finally abandoned in 1855. It was granted to the descendants of the *Bounty* mutineers, in 1857. A part of their number (about one hundred) became dissatisfied, and returned to Pitcairn. I am told that, on some parts of the island, there is a perfect network of underground workings, such as tunnels, shafts, etc., made by the prisoners, more to occupy the time of a horrible existence, than for any other purpose.

THE CHATAM GROUP.

Between latitudes 43 deg. 30 min., and 45 deg. 20 min. south, and longitudes 176 deg. 10 min., and 177 deg. 20 min. west—about six hundred miles to the east of New Zealand, and under the same rule, are the Chatam Islands. There are fifteen in the group, if we count the islets, with an area of about eight hundred square miles, and a population not exceeding five hundred. Chatam, Southeast and Pitt, are of some importance. growing all the products of temperate climes, when properly cultivated.

Through wars with the Maoris of New Zealand, the inhabitants have almost disappeared, and agriculture neglected, leaving little to be found of interest, outside of the bleak comforts of a South Sea whaling station.

The geological formation is that of New Zealand ; the soil very fertile, but without the extensive floral growth of the former. Some curious lakes and lagoons, of brackish water, are found here—often many miles in extent, and separated from the sea, at some points,

by barriers of sand, but a few hundred yards in width.
Innumerable aquatic birds make of these a favorite
resort.   Whaling, and other fisheries, form the principal interest of the group at present.   The islands were
discovered in 1791, by Lieutenant Broughton, who
named them after the vessel he commanded.

PONAPE OR ASCENSION ISLAND.

This, the principal of a group of the eastern Carolines (already briefly alluded to), lies within latitude
6 deg. 43 min. north, and longitude 158, and 158 deg.
30 min. east.   In addition to its being one of the principal stations of the Congregational Missions in the
South Sea, considerable interest has been attached to
the island, from the remains of ancient ruins, and other
evidences of a former civilization, being found there.
Of these, Captain Cheyne says:

Near Metalanien Harbor are some interesting
ruins, which are, however, involved in obscurity; the
oldest inhabitants being ignorant of their origin, and
having no tradition bearing any reference to their history.   That a fortified town once stood upon this spot,
and not built by savages, cannot be doubted; the style
of the ruins giving strong proofs of civilization.   Some
of the stones measure eight to ten feet in length, are
squared on six sides, and have evidently been brought
thither from some civilized country, there being no
stones on the island, similar to them.*   Streets are
formed in several places, and the whole town appears

---

* It has already been stated, in this work, that the material from
which former buildings, fortifications, monuments, statuary, etc., had
been constructed here, and at Strong and Easter Islands, was found in
quarries in the interior.

to have been a succession of fortified houses. Several artificial caves were also discovered within the fortifi- cations.

This town was, doubtless, at one time, the strong- hold of pirates; and, as the natives can give no account of it, it seems possible that it was built by Spanish buccaneers, some two or three centuries ago. The supposition is confirmed by the fact, that, about three or four years ago, a small brass cannon was found on one of the mountains, and taken away by H. M. S. *Larne*. Several clear places are also to be seen, a little inland, at different parts of the island, some of which are many acres in extent, clear of timber, and perfectly level. Upon one of these plains, called K-pau, near Kiti (Roan Kiddi) Harbor—and which I have fre- quently visited—is a large mound, about twenty feet wide, eight feet high, and a quarter of a mile in length. This must evidently have been thrown up for defense, or as a burial place for the dead, after some great battle. Similar ruins are to be found at Strong Island, of which the natives can give no account.

STRONG ISLAND.

Kusaie (Ualan) or Strong Island, at the eastern extremity of the Carolines, was discovered and named by Captain Crozer, an American, in 1804. It has been regarded with some interest, of late days, in the hope, that the ruins and monuments found there, might afford an explanation or clue to the origin of the ancient island races. In speaking of Pane Bay, the principal harbor, Captain Hammet describes some remains of stone architecture (also alluded to by D'Urville), which was the subject of much speculation; but Dr. Gullick

ascertained, that they were not ancient, but were built
for protection, and in some cases as monuments.´

OCEAN ISLAND.

Located in 28 deg. 22 min. north latitude, and 178
deg. 27 min. west longitude—with its surroundings of
dangerous barrier reefs, and comprised of barren sand
dunes, is unimportant, except for the dangers offered
to the navigator. It is made historical, as the scene
of the wrecks of the *Gladstone*, the American whale
ship *Parker*, and the *Saginaw*, of the United States
Navy. Another island, about one degree below the
line, and south from the Marshall group, known as
Ocean Island, with still another of the same name (the
northernmost of the Enderby group), should suggest
a change of the name of two of the islands, with a like
change on maps and charts of the Pacific, to prevent
confusion.

THE DEPTHS OF THE OCEAN.

We dive, says Schleiden, into the liquid crystal of
the Indian Ocean (a description serving alike for the
tropical Atlantic and Pacific), and it opens to us the
most wondrous enchantments, reminding us of fairy
tales in childhood's dreams. The strangely branching living thickets bear living flowers. Dense masses
of Meandrinas and Astræas, with the leafy, cup-shaped
expansions of the Explanarias, the variously ramified
Madrepores, which are now spread out like fingers,
now rise in trunk-like branches, and now display the
most elegant array of interlacing branches. The coloring surpasses everything—vivid green alternates
with brown or yellow; rich tints of purple, from pale
red-brown to the deepest blue: brilliant rosy, yellow

or peach-colored Nullipores overgrow the decaying
masses, and are themselves interwoven with the
pearl-colored plates of the Reptipores, resembling
the most delicate ivory carvings.   Close by, wave the
yellow and lilac fans, perforated like trellis-work of
the Gorgonias. The clear sand of the bottom is
covered with the thousand strange forms and tints
of the sea-urchins and star-fishes. The leaf-like flus-
tras and escharas adhere like mosses and lichens to
the branches of the corals; the yellow, green and
purple striped limpets cling like monstrous cochineal
insects upon their trunks. Like gigantic cactus-blos-
soms, sparkling in the most ardent colors, the sea-
anemones, expand their crowns of tentacles upon the
broken rocks, or more modestly embellish the flat
bottom, looking like beds of variegated ranunculuses.
Around the blossoms of the coral shrubs play the
humming-birds of the ocean, little fishes sparkling
with red or blue metallic glitter, or gleaming in golden
green, or in the brightest silvery luster. Softly, like
spirits of the deep, the delicate milk-white or bluish
bells of the jelly-fishes float through this charmed
world. Here, the gleaming violet and gold-green
Isabelle, and the flaming yellow, black and vermillion
striped coquette chase their prey; there, the band-
fish shoots snake-like through the thicket, like a long
silk ribbon, glittering with rosy and azure hues.
Then comes the fabulous cuttle-fish, decked in all
colors of the rainbow, but marked by no definite outline, appearing and disappearing, intercrossing, join-
ing company and parting again, in most fantastic
ways; and all this in the most rapid change, and amid
the most wonderful play of light and shade, altered
by every breath of wind and every slight curling of

the surface of the ocean. When day declines, and the shades of night lay hold upon the deep, this fantastic garden is lighted up in new splendor. Millions of glowing sparks, little microscopic medusas and crustaceons, dance like glow-worms through the gloom. The sea-feather, which by daylight is vermillion-colored, waves in a greenish, phosphorescent light. Every corner of it is lustrous. Parts which by day were dull and brown, and retreated from sight, amid the universal brilliancy of color, are now radiant in the most wonderful play of green, yellow and red light; and, to complete the wonders of the enchanted night, the silver disc, six feet across, of the moon-fish, moves, slightly luminous, among the cloud of little sparkling stars.

The most luxuriant vegetation of a tropical landscape cannot unfold as great wealth of form, while in the variety and splendor of color it would stand far behind this garden landscape, which is strangely composed exclusively of animals, and not of plants; for, characteristic as the luxuriant development of vegetation of the temperate zones is of the sea-bottom, the fullness and multiplicity of the marine Fauna is just as prominent in the regions of the tropics. Whatever is beautiful, wondrous or uncommon in the great classes of fish and Echinoderms, Jelly-fishes and Polypes, and the Mollusks of all kinds, is crowded into the warm and crystal waters of the tropical ocean, rests in the white sands, clothes the rough cliffs, clings where the room is already occupied, like a parasite, upon the first comers, or swims through the shallows and depths of the elements—while the mass of the vegetation is of a far inferior magnitude. It is peculiar in relation to this that the law valid on land, ac-

cording to which the animal kingdom being better
adapted to accommodate itself to outward circumstances, has a greater diffusion than the vegetable
kingdom—for the Polar Seas swarm with whales,
seals, sea-birds, fishes and countless numbers of the
lower animals, even where every trace of vegetation
has long vanished in the eternally frozen ice, and
the cooled sea fosters no sea-weed—that this law, I
say, holds good also for the sea, in the direction of
its depth; for when we descend, vegetable life van-
ishes much sooner than the animal, and even from the
depths to which no ray of light is capable of penetrating, the sounding-lead brings up news at least of
living infusoria.

In concluding with Schleiden's description of the
shallower depths, it might be well to add something
on the characteristics of deeper soundings.

According to the records published of the voyage
of the *Challenger*, in 1872–3, after leaving the Admiralty Islands, on the 10th of March, a course was
shaped for Yokohama, with the intention of reaching
Guam, one of the Ladrone Islands. They lost the
trades in latitude 17 deg. north, and after that, had a
succession of easterly, northeasterly, and baffling
winds from every point of the compass, except where
it was wanted; thus preventing their visiting either the
Carolines or Ladrones, which were passed some one
hundred miles to leeward. On the 23d of March, in
latitude 11 deg. 24 min. north, and longitude 143 deg.
16 min. east, bottom was touched at 4.475 fathoms—
the deepest successful soundings made during the
whole cruise. Specimens from that depth showed a
dark, volcanic sand, mixed with manganese. In consequence of the enormous pressure at that depth

(some five tons on the square inch), most of the thermometers were crushed. However, one stood the test, and showed a temperature of 33.9 deg., the surface temperature being 80 deg. Three other attempts were made to determine the temperature of water at these great depths, but in every instance the instruments came to the surface in a damaged condition. In the case of the *Challenger* soundings, already noted above, the pressure would be fully six tons per square inch, at a depth of 4,475 fathoms, or nearly five and one-fifth miles. At other points of the Pacific—one, in particular, 350 miles east from Yeddo, Japan—a depth was obtained of 3,950 fathoms—not quite four and one-half miles. In the Torres Straits (separating Australia from New Guinea), 2,650 fathoms was shown, being 160 fathoms over three miles. Between New Zealand, the Tongas and Fiji Islands, 1,100 to 2,900 fathoms was found, or from one and a quarter to over three and a quarter miles. Still deeper soundings have been taken, recently, in the different oceans, but in exceptional cases only. In the Atlantic, 90 miles north of the island of St. Thomas, 3,875 fathoms, and another, near St. Helena, 4,500 fathoms of line was paid out, before the bottom was reached. At St. Thomas, the bulbs of the thermometers, constructed to sustain a pressure of three tons to the square inch, were crushed like egg-shells. The temperature of the water, generally, in deep soundings, is below the freezing-point, and life is found only in its primal forms.

The immense depths reached (but a few of which I have recorded), are the results of practical tests, and are not theoretical. In all tests of this character, it is absolutely necessary to bring up samples from the

floor of the sea; otherwise, strong under-currents, and
the pressure of the water, may give a depth, not altogether reliable.   Off the west coast of South America,
a very deep sounding was obtained, that was more
amusing than reliable.   Such a length of cable was paid
out (about ten miles), being carried by the under-cur-
rent in one direction, and by the surface flow in another, that the force exerted in hauling in, broke the
line.

BOTTOM OF THE SEA.

Of the great ocean's floor—a deep vale, majestic
and immense in area, lying miles below the level of
our present shore lines—it might be said, that, if the
water could be taken away from the great basin of the
Pacific, not many centuries would elapse, before its
floor would resemble other portions of the land.
Then, if it were possible for the human vision to encompass the scene, the valleys, plains, deserts, the
mountain chains and ravines, the hills and glades,
the stately course of rivers, or the meandering of
brooks, would, like a vast panorama, enchant the
view.   The plumed tufts of the cocoanut tree, or its
northern neighbor, the pine, would wave in the breeze,
or bend in the storm.   The present home, of the
leviathan of the deep, and the busy little coral insect,
would give place to other forms of life, and the island
world would fade from view, like the slow awakening
from a summer's dream.

www.ingramcontent.com/pod-product-compliance
Lightning Source LLC
Chambersburg PA
CBHW030905270326
41929CB00008B/589